T0310588

The Mathematics of
Open Quantum Systems

**Dissipative and Non-Unitary Representations
and Quantum Measurements**

The Mathematics of Open Quantum Systems

Dissipative and Non-Unitary Representations and Quantum Measurements

Konstantin A Makarov

University of Missouri, USA

Eduard Tsekanovskii

Niagara University, USA

 **World Scientific**

NEW JERSEY · LONDON · SINGAPORE · BEIJING · SHANGHAI · HONG KONG · TAIPEI · CHENNAI · TOKYO

Published by

World Scientific Publishing Co. Pte. Ltd.

5 Toh Tuck Link, Singapore 596224

USA office: 27 Warren Street, Suite 401-402, Hackensack, NJ 07601

UK office: 57 Shelton Street, Covent Garden, London WC2H 9HE

Library of Congress Control Number: 2021059101

British Library Cataloguing-in-Publication Data
A catalogue record for this book is available from the British Library.

THE MATHEMATICS OF OPEN QUANTUM SYSTEMS
Dissipative and Non-Unitary Representations and Quantum Measurements

ISBN 978-981-124-122-2 (hardcover)
ISBN 978-981-124-123-9 (ebook for institutions)
ISBN 978-981-124-124-6 (ebook for individuals)

For any available supplementary material, please visit
https://www.worldscientific.com/worldscibooks/10.1142/12395#t=suppl

Typeset by Stallion Press
Email: enquiries@stallionpress.com

Printed in Singapore

In respectful memory of our beloved teachers M. Livšic and B. Pavlov.

This book is dedicated to the memory of the remarkable Human Beings and Mathematicians Michail Samoilovich Livšic (M.S.) and Boris Sergeevich Pavlov (B.S.). Their pioneering research in the theory of non-self-adjoint operators and applications to scattering problems and system theory have attracted many researchers and made the present book possible.

The Apostolic service to students, colleagues and the mathematical community provided by M.S. and B.S. was enormous and their good deeds will never be forgotten. The light of scientific accomplishments of M.S. and B.S. shine brightly and is succinctly described by the poetic word of Galina Volchek:

> Уходя оставьте Свет! Это больше, чем остаться...
> Это лучше, чем прощаться и важней, чем дать совет...
> Уходя оставьте Свет - перед ним отступит холод!
> Свет собой заполнит город... Даже если вас там нет...

ACKNOWLEDGMENTS

We are very grateful to M. Ashbaugh, W. Banks and S. Belyi for the invaluable help in the preparation of this monograph for publication. K.A.M. is indebted to A. B. Plachenov for stimulating discussions. The authors are also grateful to the three referees for their valuable comments and suggestions.

K.A.M. is grateful to his wife Marina and son Konstantin for their patience and continuous support.

E.T. is very thankful to his son Vladislav, daughter-in-law Elena, and granddaughters H and R for overwhelming continuous help, support and care.

Research of K.A.M. was partially supported by the Simons collaboration grant 00061759.

PREFACE

The main goal of this monograph is to develop a mathematical framework that accommodates an adequate description of the results of continuous observation of quantum systems. It is well known that a quantum observation/measurement always affects the system subject to it, and therefore the system can no longer be considered completely isolated. Instead, it should be treated as part of a more general system in which the presence of an observer/measuring device is taken into account. As such, the initial system should be regarded as an open quantum system that interacts with a part of the larger system.

In the abstract setting, dealing with open systems assumes the presence of communication channels through which the interaction is carried out, both between parts of the system in question and with the outside world. A special case of open systems is the class of dissipative systems, the dynamics of which are governed by a strongly continuous semigroup of contractions. By applying the canonical dilation procedure, such open systems can always be viewed as a part of a larger closed system: the resulting Hamiltonian of the dilated system is chosen to be the self-adjoint dilatation of the generator of the semigroup, while the space of (pure) states of the larger system can be identified with the extended Hilbert space where the dilated operator has been realized as a self-adjoint operator. Despite its mathematical attractiveness, this dilation method is incompatible with the physical requirement that the total energy of the obtained large isolated system must be bounded from below.

The transition in the opposite direction is usually associated with the reduction of the unitary evolution onto a subspace, in most cases lacks the semigroup property and needs a special consideration. The situation changes dramatically if the reduced description of the evolution is accompanied by continuous monitoring of the system. Under certain circumstances, the exponential decay of the states under continuous monitoring can be justified even if the energy distribution of the state is semi-bounded from below and the spectrum of the system is discrete. Therefore, within the continuous monitoring paradigm, one can bypass applying the Weisskopf-Wigner method that can only give an approximate description of the decay processes only. In the same time, the exponential decay under the continuous monitoring scenario fills in the gap between the quantum Zeno effect (Turing's paradox) and the anti-Zeno phenomenon, and what is also important is, it gives a fresh look at the descent of dissipative operators and opens up new perspectives for their applications in quantum theory.

In the present monograph we discuss two closely related topics. In the first part, based on the study of unitary invariants of symmetric operators, we provide the complete classification of the simplest dissipative solutions of the Heisenberg commutation relations (in the Weyl form). The second part of the monograph deals with mathematical problems of continuous monitoring of general quantum systems initially prepared in a pure state. Special attention is paid to the discussion of the behavior of massless particles on a ring under continuous monitoring.

CONTENTS

One might still like to ask: "How does it work? What is the machinery behind the law?" No one has found any machinery behind the law. No one can "explain" any more than we have just "explained." No one will give you any deeper representation of the situation. We have no ideas about a more basic mechanism from which these results can be deduced.

Richard Feynman, *The Feynman Lectures on Physics, Volume III*

Part 1

Representations of Operator Commutation Relations

Chapter 1

INTRODUCTION

The classical Stone-von Neumann theorem [27, 83, 125, 127] states that the unitary representations of the canonical commutation relations (CCR) of Quantum Mechanics in the Weyl form [139]

$$U_t V_s = e^{ist} V_s U_t \tag{1.1}$$

for strongly continuous, one-parameter groups of unitary operators U_t and V_s in a separable Hilbert space $\mathcal{H}$ are unitarily equivalent to a direct sum of copies of the unique irreducible system in the Hilbert space $\mathcal{H} = L^2(\mathbb{R})$ with

$$(U_t f)(x) = \exp(ixt) f(x) \quad \text{and} \quad (V_s f)(x) = f(x - s).$$

For the history of the subject we refer to [115] where one can find a thorough discussion of the further generalizations initiated by G. W. Mackey in his ground-breaking paper [83], and the subsequent development in number theory due to A. Weil [137]. We refer also to the series of publications [13, 21, 38, 57, 72, 81, 113, 114] where the interested reader can find a truly extensive body of information on the subject.

The CCR (1.1) can be reformulated in an equivalent infinitesimal form (see [27], [125]) as a relation for the self-adjoint generator A of the group $V_s = e^{isA}$

$$U_t A U_t^* = A + tI \quad \text{on Dom}(A), \quad t \in \mathbb{R}, \tag{1.2}$$

or as the equality invoking the spectral measure $E(d\lambda)$ of the self-adjoint operator A

$$U_t E(\delta) U_t^* = E(\delta - t), \ t \in \mathbb{R}, \quad \delta \text{ a Borel set.}$$

It is worth mentioning that rewriting relations (1.1) in its infinitesimal (semi-Weyl) form (1.2) opens a way for the further developments and generalizations in various directions.

3

In this book we choose a presentation line taking into account the following observation: the Stone-von Neumann uniqueness result [98] implies that if a self-adjoint operator A satisfies (1.2), then A always admits a symmetric restriction $\dot{A} \subset A$ with deficiency indices $(1,1)$ such that the same commutation relations

$$U_t \dot{A} U_t^* = \dot{A} + tI \quad \text{on} \quad \text{Dom}(\dot{A}), \quad t \in \mathbb{R}, \tag{1.3}$$

hold.

Given the commutation relations for a symmetric operator (1.3), see Hypothesis 3.1, the following natural problems can be posed:

(I) *a) Characterize such symmetric operator solutions $\dot{A}$ up to unitary equivalence;*

     b) Provide an intrinsic characterization of those solutions.

(II) *Find the maximal dissipative solutions $\widehat{A}$ to the infinitesimal Weyl relations of the form*

$$U_t \widehat{A} U_t^* = \widehat{A} + tI \quad \text{on} \quad \text{Dom}(\widehat{A}) \tag{1.4}$$

such that $\dot{A} \subset \widehat{A} \subset (\dot{A})^*$.

Problem (I) b) was posed in [52] and we will refer to it as the *Jørgensen-Muhly problem.*

Notice that in this situation the semigroup $V_s = e^{is\widehat{A}}$, $s \geq 0$, generated by the dissipative operator $\widehat{A}$ and the unitary group $U_t = e^{itB}$, generated by a self-adjoint operator $B = B^*$, satisfy the restricted Weyl commutation relations

$$U_t V_s = e^{ist} V_s U_t, \quad t \in \mathbb{R}, \quad s \geq 0. \tag{1.5}$$

More generally, one can ask to provide the complete classification (up to mutual unitary equivalence) of the pairs of corresponding generators $(\widehat{A}, B)$ under the solely assumption that the generator $\widehat{A}$ is an extension of a symmetric operator with arbitrary deficiency indices (m, n).

Much progress has been achieved in this area of research (see [9, 50, 51, 52, 53, 54, 55, 121, 122], also see [88]). For instance, it is known that a semi-group satisfies the restricted Weyl relations if and only if the characteristic function of its generator has a particularly simple form [55, Theorem 20]. Moreover, in this case, the restricted Weyl system can be dilated to a canonical Weyl system in an extended Hilbert space [55, Theorem 15]. However, to the best of our knowledge, the *complete classification* of irreducible representations of the restricted commutation relations (1.5), even in the simplest case of deficiency indices $(1,1)$, has not been obtained yet.

In this book we give the *complete solution to problem* (II) under the assumption that the generator $\widehat{A}$ of the semi-group V_s is a dissipative quasi-selfadjoint extension [3] of a prime symmetric operator $\dot{A}$ with deficiency indices $(1,1)$. As in the Stone-von Neumann theorem, we show that the pair $(\widehat{A}, B)$ of generators is mutually unitarily equivalent to the "canonical" pair $(\widehat{\mathcal{P}}, \mathcal{Q})$ on a metric graph $\mathbb{Y}$, finite or infinite. Here $\widehat{\mathcal{P}}$ stands for a dissipative differentiation (*momentum*) operator on $\mathbb{Y}$ with appropriate vertex boundary conditions and $\mathcal{Q}$ is the self-adjoint multiplication (*position*) operator on the graph $\mathbb{Y}$. In contrast to the Stone-von Neumann uniqueness theorem, where the corresponding graph is just the real axis (with no reference vertices), the graph geometry of $\mathbb{Y}$ is more varied (see Definition 13.6 for the classification). Moreover, the knowledge of the complete set of unitary invariants of the solutions to the commutation relations determines not only the geometry of the metric graph $\mathbb{Y}$ but also the location of the central vertex of the graph. For instance, given a solution of the commutation relations on a metric graph, one obtains new series of unitarily inequivalent solutions by shifting the graph.

Our approach is based on the detailed study of unitary invariants of operators such as the Livšic and/or Weyl-Titchmarsh functions associated with the pair of a symmetric operator and its self-adjoint (reference) extension as well as the characteristic function of a dissipative triple of operators. A comprehensive study of the concept of a characteristic function associated with various classes of non-selfadjoint operators, in particular, with applications to scattering theory, system theory and boundary value problems one can find in [10, 72, 73, 77, 79, 82, 95, 100, 105, 106] as well as in [1, 4, 7, 14, 15, 17, 18, 19, 24, 25, 68, 69, 78, 80, 81, 94, 96, 97, 99, 104, 107, 111, 118, 128, 129, 134, 140, 142].

The departure point for our study of commutation relations is structure Theorem 3.5. This result states that in the situation in question the characteristic function of a dissipative triple is either (i) a constant, or (ii) a singular inner function in the upper half-plane with "mass at infinity", or (iii) the product of those two. The examples of differentiation operators (more precisely, a triple of those) with either a constant or entire characteristic function are known (see, e.g., [3, Ch. IX]). The construction of the model differentiation operator/triple in the general case (iii) can be achieved in the framework of operator coupling theory [87]. Notice that those examples of differentiation operators are *the building blocks* in our approach. In particular, addressing Problem (I) a), we obtain the complete classification of the symmetric operators with deficiency indices $(1,1)$ that solve the infinitesimal relations (1.3) up to unitary equivalence. We also provide an intrinsic characterization of the corresponding symmetric operator solutions to the commutation relations, thus giving a *comprehensive answer to the Jørgensen-Muhly problem* (I) *b*) [52] (see Remark 12.3).

In the second part of the book we address the problem of how a closed quantum system becomes an open one under continuous monitoring. We refer to a selected list of monographs [11, 12, 17, 27, 29, 46, 74, 75] where different aspects of open/closed quantum systems theory are discussed. In addition, within the continuous monitoring paradigm, we study in detail theoretical foundations for complementarity of the *Quantum Zeno* and *Exponential Decay* scenarios in frequent quantum measurements experiments. Notice that considering the corresponding open dissipative quantum system as part of a larger closed system within the self-adjoint dilation scheme generates solutions to the canonical commutation relations (see Section 15.5 for more details). For the relevant background material, we refer to [6, 58, 93] and [36, 66, 138], respectively, and the references therein. In this context, we also want to mention the revolutionary paper by Gamow [39] who was the first to introduce quantum states with "complex" energies and, based on this concept, gave an explanation for the decay law for a quasi-stationary state.

In the framework of our formalism we give a justification for the exponential decay scenario (under continuous monitoring) by recognizing the phenomenon as a variant of the Gnedenko-Kolmogorov 1-stable limit theorem. Having this link in mind, we obtain several principal results in quantum measurement theory. In particular, we show that a "typical smooth" state of a material (massive) particle under continuous monitoring is either a *Zeno state or an anti-Zeno state* (see Theorems 16.3 and 16.4). In contrast to that, for the systems of massless particles (fields) the situation is quite different: if the Hamiltonian of the system is given by the first order differentiation operator, then the *quantum Zeno and exponential decay scenarios are complementary* instead. In addition, it turns out that for that kind of systems, the decay rate is rather sensitive to the choice of a self-adjoint realization of the Hamiltonian on the metric graph, especially if the graph is not simply connected. From the point of view of physics, this phenomenon is a manifestation of the Aharonov-Bohm effect. That is, in the absence of the magnetic field, the magnetic potential by itself affects the magnitude of the decay rate in this case (see Theorem 17.4 and Chapter 18, eq. (20.14)). We also notice that the existence of states the decay rate of which is independent of the Aharonov-Bohm field is closely related to the search for dissipative operator solutions of the infinitesimal Weyl commutation relations (1.4).

As an illustration, within continuous monitoring paradigm we discuss a *Gedankenexperiment* where the renowned *exclusive* and *interference* measurement alternatives in quantum theory can be rigorously analyzed. In addition, on the basis of an explicitly soluble model, we present a variant of the celebrated "double-slit experiment" in a way that is accessible for mathematicians (see Theorem 20.1, eq. (20.3), and Theorem 21.1, eq. (21.3)).

We conclude our treatise by the discussion of limit theorems in the framework of operator coupling theory. More specifically, we introduce a new mode of convergence for dissipative operators (in distribution) and show that basic dissipative solutions to the commutation relations can be considered analogs of stable distributions in the orthodox probability theory. This observation, in our opinion, sheds some light on the foundations of dissipative quantum theory of open systems which our dear teachers M.S. and B.S. dedicated their scientific life to.

The book is organized as follows.

• In Chapter 2, following our work [86], we recall the concept of the Weyl-Titchmarsh and the Livšic functions associated with the pair of operators $(\dot{A}, A)$ where $\dot{A}$ is a symmetric operator with deficiency indices $(1, 1)$ and A its self-adjoint extension. Given a dissipative quasi-selfadjoint extension $\widehat{A}$ of $\dot{A}$, we also introduce a characteristic function associated with the triple $(\dot{A}, \widehat{A}, A)$ and provide a characterization of the projection of the deficiency subspace $\mathrm{Ker}((\dot{A})^* - zI)$ onto the subspace $\mathrm{Ker}((\dot{A})^* - \bar{z}I)$ along the domain of the dissipative operator $\widehat{A}$ in terms of the characteristic function of the triple (see Proposition 2.2).

• In Chapter 3 we study symmetric operators with deficiency indices $(1, 1)$ that satisfy the semi-Weyl commutation relations (1.3) (see Hypothesis 3.1). We show that if $\dot{A}$ admits a self-adjoint extension A that solves the same commutation relations as the operator $\dot{A}$ does, then the Livšic function associated with the pair $(\dot{A}, A)$ is identically zero in the upper half-plane (see Theorem 3.3). In this case the corresponding Weyl-Titchmarsh function is z-independent and coincides with $i = \sqrt{-1}$ on the whole upper half-plane. On the other hand, if $\dot{A}$ has no self-adjoint extension that solves the same commutation relations but does have a dissipative extension solving (1.4), then Theorem 3.5 asserts that the characteristic function of the corresponding triple is periodic with a real period and has a particularly simple form. Notice that in the case where $\dot{A}$ admits both a self-adjoint A and a dissipative extension $\widehat{A}$ that solve (1.4), the characteristic function of the corresponding triple $(\dot{A}, \widehat{A}, A)$ is a constant from the open unit disk.

As a corollary to Theorems 3.3 and 3.5 one gets an explicit representation for the Livšic function of an arbitrary symmetric operator satisfying Hypothesis 3.1 (see Corollary 3.7 for a precise statement).

• In Chapter 4 we discuss first order symmetric differential operators on a metric graph $\mathbb{Y}$ assuming that the graph is in one of the following three cases: the graph (i) is the real axis with a reference point, (ii) is a finite interval, and finally, (iii) is obtained by attaching a finite interval to the real axis. Notice that the symmetric operators in question solve the semi-Weyl commutation relations (1.3) (see Remark 4.6).

• In Chapter 5 we give a useful parameterization for the family of all self-adjoint as well as quasi-selfadjoint extensions of a symmetric differentiation discussed in Chapter 4. We also show that any self-adjoint realization D_Θ of the symmetric differentiation operator $\dot{D}$ on the graph $\mathbb{Y}$ in Case (iii) serves as a minimal self-adjoint dilation of an appropriate quasi-selfadjoint dissipative differentiation operator on the metric graph in Case (ii) (see Theorem 5.7).

• In Chapter 6 we examine the Livšic functions associated with the pair $(\dot{D}, D_\Theta)$ where $\dot{D}$ is the symmetric operator on the metric graph $\mathbb{Y}$ (in Cases (i)–(iii)) and D_Θ is its arbitrary reference self-adjoint extension. In particular, we provide a complete solution of Problem (I) $a)$ (see Corollary 6.4).

• Chapter 7 is devoted to the comprehensive study of the corresponding Weyl-Titchmarsh functions.

• In Chapter 8, given the graph $\mathbb{Y}$ in one of the Cases (i)–(iii), we introduce maximal dissipative differentiation (model) operators the domains of which are invariant with respect to the group of gauge transformations. Those are the prototypes of general dissipative solutions to the commutation relations (1.4). Notice that in Cases (i) and (iii) the corresponding boundary conditions are determined not only by the geometry of the metric graph $\mathbb{Y}$ but also by its peculiar "conductivity" exponent k, $0 \leq k < 1$, which we call the *quantum gate coefficient*. We also give an explicit informal description of the associated contraction semi-groups generated by these operators. A visualization of the corresponding dissipative dynamics is shown on Figures **1–3**.

• In Chapter 9 we evaluate the characteristic function of the triples associated with the model dissipative operators discussed in Chapter 8 and prove the converse of important structure Theorem 3.5 (see Theorem 9.7).

• In Chapter 10, following the line of research initiated by Livšic [78], who was the first to discover the connection between the Heisenberg scattering matrix and the characteristic function of a dissipative operator (also see [1, 14, 72]), we focus on the discussion of general quasi-selfadjoint dissipative differentiation operators $\widehat{d}$ on the metric graph $\mathbb{Y}$ in Case (ii). In particular, we relate the characteristic function of the corresponding triple with the reciprocal of the transmission coefficient in a scattering problem for a self-adjoint dilation of $\widehat{d}$, the magnetic Hamiltonian. Using the fundamental relation (C.1) in Appendix C between the characteristic and Livšic functions, we also get a representation for the transmission coefficient via the Livšic function and the von Neumann parameter of the triple (see Corollary 10.4 combined with eq. (10.3)). Notice that both the differentiation operator $\widehat{d}$ and the magnetic Hamiltonian solve commutation relations 1.4 with respect to a discrete subgroup of the unitary group U_t.

• In Chapter 11 we discuss uniqueness results for symmetric operators that commute with a unitary. In particular, we show that the unitary group U_t from (1.4) is uniquely determined (up to a character) unless the dissipative solution to the commutation relations has point spectrum filling in the whole upper half-plane. In the latter (exceptional) case the representation $t \mapsto U_t$ is reducible and splits into the orthogonal sum of two irreducible representations uniquely determined up to a unitary character.

• In Chapter 12 we characterize maximal dissipative solutions to the semi-Weyl commutation relations (1.4) that extend a prime symmetric operator satisfying Hypothesis 3.1 (see Theorem 12.1). As a by-product of these considerations, we provide an intrinsic characterization of these symmetric operators thus solving the Jørgensen-Muhly problem in the particular case of the deficiency indices $(1, 1)$.

• In Chapter 13 we study solutions to the restricted Weyl commutation relations (1.5) for a unitary group $U_t = e^{iBt}$ and a semi-group $\widehat{V}_s = e^{i\widehat{A}s}$ of contractions. Under the assumption that the generator $\widehat{A}$ of $\widehat{V}_s$ is a quasi-selfadjoint extension of a prime symmetric operator with deficiency indices $(1, 1)$ we characterize the pairs of generators $(\widehat{A}, B)$ up to mutual unitary equivalence. In particular, we show that the generators $\widehat{A}$ and B can be realized as the dissipative differentiation operator $\widehat{\mathcal{P}}$ on the metric graph $\mathbb{Y}$ in one of the Cases I*, I–III (see Definition 13.6 for the classification) with appropriate boundary conditions at the vertex of the graph and the multiplication operator $\mathcal{Q}$ on the graph, respectively (see Theorem 13.12). In contrast to the classic Stone-von Neumann uniqueness result, the pairs $(\widehat{\mathcal{P}}, \mathcal{Q})$ are not unitarily equivalent for different choices of the center of the graph. However, the uniqueness theorem in the self-adjoint case can be adopted to fit the format of its non-selfadjoint counterpart (see Theorem 13.14). In particular, we get the *full version* of the Stone-von Neumann Theorem (see Corollary 13.16).

• In Chapter 14 we consider a family of unitary solutions to the canonical Weyl commutation relations (1.1) on the full metric graph $\mathbb{X}$ obtained by composing two identical copies of the metric graph $(-\infty, \mu) \sqcup (\mu, \infty)$, $\mu \in \mathbb{R}$ is a parameter. The corresponding self-adjoint momentum differentiation operator $\mathcal{P}$, the generator of the group V_s of shifts, is determined by the boundary conditions at the vertex μ with the bond S-matrix given by

$$S = \begin{pmatrix} k & -\sqrt{1-k^2} \\ \sqrt{1-k^2} & k \end{pmatrix} \quad \text{for some } 0 \leq k < 1,$$

and the generator $\mathcal{Q}$ of the second group U_t is just the position operator on the graph $\mathbb{X}$. As long as the generator $\widehat{A}$ of $\widehat{V}_s$ is a quasi-selfadjoint extension of a prime symmetric operator with deficiency indices $(1,1)$, the structure Theorem 14.1 shows that up to mutual unitary equivalence any solution to the restricted Weyl commutation relations (1.5) can be obtained by an appropriate compression of the unitary groups U_t and V_s onto some subspace $\mathcal{K}$ of the Hilbert space $L^2(\mathbb{X})$. In fact, the subspace $\mathcal{K}$ coincides with the Hilbert space $L^2(\mathbb{Y})$ where $\mathbb{Y} \subset \mathbb{X}$ is a subgraph of $\mathbb{X}$ in one of the three canonical cases discussed above. Notice that the subspace $\mathcal{K}$ reduces the multiplication group $U_t = e^{it\mathcal{Q}}$ and is coinvariant for the group of shifts $V_s = e^{is\mathcal{P}}$. For a pictorial description of the corresponding unitary dynamics on the full metric graph $\mathbb{X}$ we refer to Figure **4**.

In the second part of the book we discuss applications to decay phenomena in quantum systems theory.

• In Chapter 15 we recall the concept of continuous monitoring of a quantum system and describe possible *ex-post* monitoring scenarios: the Quantum Zeno and Anti-Zeno effects as well as the Exponential Decay phenomenon in frequent measurements theory. We also provide an example of unstable (pure) states of the quantum oscillator that decay exponentially under continuous monitoring of the system which eventually confirms the conclusions of the phenomenological Weisskopf-Wigner theory of decay [138].

• In Chapter 16 we discuss the quantum Zeno versus Anti-Zeno effect alternative for massive particles. Applying 1/2- and 3/2-stable limit theorems we also show that continuous monitoring in nonlinear time scales leads to exponential decay of some appropriate states in a quantum systems the Hamiltonian of which is given by the free Schödinger operator on the semi-axis with (various) Robin type boundary conditions at the origin (see Theorems 16.3 and 16.4).

• In Chapter 17 we examine the exponential decay phenomenon for "zero-mass" systems where the exponential decay typically alternates with the quantum Zeno scenario (see Theorem 17.4).

• In Chapter 18 we recall main concepts for the "exclusive" versus "interference" alternatives theory going back to the celebrated two slit experiment in quantum mechanics.

• In Chapter 19 we consider a model of a quantum system on a ring that describes a motion of a (relativistic) massless particle and set the stage for monitoring such systems within the continuous observation paradigm.

• In Chapter 20 we show that in some cases continuous monitoring of the model quantum system triggers emission of particle, see Theorem 20.1. For instance, this phenomenon occurs if the initial state has a unique jump-point discontinuity on the ring. The magnitude of the state decay in this case can be theoretically predicted as if the quantum particle were a wave. That is, the particle interferes with itself at the point of observations (where the wave functions has a jump) and the results of this interference can informally be explained in the framework of the "interference" alternatives theory. We also illustrate some features in the state decay under continuous monitoring of a massless particle moving along the Aharonov-Bohm ring (see Corollary 20.2 and eq. (20.14), cf. [59, 67]).

• In Chapter 21 we discuss results of continuous monitoring of open quantum systems on a ring under the hypothesis that the time evolution of the system is governed by a semi-group of contractions. If the initial state of the system satisfies the radiation condition (21.4), that is, it belongs to the domain of the evolution generator, then the decay rate can easily be computed using purely classical considerations, as if the quantum particle were a classical particle (see Theorem 21.1, Corollary 21.2 and the related discussion).

• In Chapter 22 we explicitly describe the self-adjoint dilation of the dissipative generator of the open system on the ring.

• In Chapter 23 we consider more general states of the open quantum system on the ring. The main result (see Theorem 23.1) states that the emission rate splits into the sum of two terms. One of the terms is due to the interference of the particle with itself at the point of observations. The second source of emission is caused by inelastic collision of the quantum particle with the point "defect" (membrane) at the observation point where the corpuscular nature of the quantum particle is fully manifested.

• In Chapter 24 we introduce the concept of convergence of dissipative operators in distribution and prove several limit theorems with respect to multiple coupling of an operator with itself. We also show that the generator of the nilpotent semi-group, one of the building blocks of the restricted commutation relations theory, can be obtained as the limit of appropriately normalized n-fold couplings of almost arbitrary dissipative operator (see Theorem 24.3).

The book has eight appendices where, for the reader's convenience, one can find relevant information scattered in the literature. Some of the results presented there are new.

In Appendix A we recall the notion of the Weyl-Titchmarsh functions as well as a characteristic function for rank-one dissipative perturbations of a self-adjoint operator. We also provide the corresponding uniqueness result (see Theorem A.5).

In Appendix B we collect necessary background material from the theory of symmetric operators.

In Appendix C we present a functional model for triples of operators following our work [86] (also see [4, Ch. 10, Sec. 10.4, p. 357] and [68, 89, 116, 117, 133]).

Appendix D contains a discussion aimed at the spectral analysis of model dissipative triple of operators.

In Appendix E we study the dependence of the Weyl-Titchmarsh, Livšic, and characteristic function under affine transformation of the operators.

In Appendix F we discuss the invariance principle for affine transformations of a dissipative operator.

In Appendix G we recall the concept of an operator coupling of two dissipative operators and discuss the corresponding multiplication theorem.

In Appendix H one can find a brief discussion of stable laws in probability theory and the formulation of the general Gnedenko-Kolmogorov limit theorem.

Some words about notation:

The domain of a linear operator K is denoted by $\mathrm{Dom}(K)$, its range by $\mathrm{Ran}(K)$, and its kernel by $\mathrm{Ker}(K)$. The restriction of K to a given subset $\mathcal{C}$ of $\mathrm{Dom}(K)$ is written as $K|_{\mathcal{C}}$. We write $\rho(K)$ for the resolvent set of a closed operator K on a Hilbert space, and K^* stands for the adjoint operator of K if K is densely defined.

Chapter 2

PRELIMINARIES AND BASIC DEFINITIONS

Let $\dot{A}$ be a densely defined symmetric operator with deficiency indices $(1,1)$ and A its self-adjoint (reference) extension.

Following [23, 41, 43, 73, 86] recall the concept of the Weyl-Titchmarsh and Livšic functions associated with the pair $(\dot{A}, A)$.

Suppose that (normalized) deficiency elements $g_\pm \in \mathrm{Ker}((\dot{A})^* \mp iI)$, $\|g_\pm\| = 1$, are chosen in such a way that

$$g_+ - g_- \in \mathrm{Dom}(A). \tag{2.1}$$

Consider the *Weyl-Titchmarsh function*

$$M(z) = ((Az + I)(A - zI)^{-1}g_+, g_+), \quad z \in \mathbb{C}_+, \tag{2.2}$$

and the *Livšic function*

$$s(z) = \frac{z - i}{z + i} \cdot \frac{(g_z, g_-)}{(g_z, g_+)}, \quad z \in \mathbb{C}_+, \tag{2.3}$$

associated with the pair $(\dot{A}, A)$. Here $g_z \in \mathrm{Ker}((\dot{A})^* - zI)$, $z \in \mathbb{C}_+$. Paying tribute to historical justice it is worth mentioning that the function $M(z)$ has been introduced by Donoghue in [23]. However, as one can see from [41, eq. (5.42)], it is elementary to express $M(z)$ in terms of the classical Weyl-Titchmarsh function which explains the terminology we use.

Clearly, the Weyl-Titchmarsh function $M(z)$ does not depend on the concrete choice of the normalized deficiency element g_+. We also remark that if

$$g'_\pm \in \mathrm{Ker}((\dot{A})^* \mp iI), \quad \|g'_\pm\| = 1,$$

is any deficiency elements such that

$$g'_+ - g'_- \in \mathrm{Dom}(A),$$

then necessarily $g'_\pm = \Theta g_\pm$ for some unimodular factor Θ. Therefore, from (2.3) it follows that the Livšic function does not depend on the choice of the deficiency elements $g_\pm$ (whenever (2.1) holds). However it may and in most of the cases does depend on the reference operator A. As far as the Weyl-Titchmarsh functions are concerned we also refer to the related concept of a Q-function introduced in [64] and discussed in [65].

Recall the important relationship between the Weyl-Titchmarsh and Livšic functions [86]

$$s(z) = \frac{M(z) - i}{M(z) + i}, \quad z \in \mathbb{C}_+. \tag{2.4}$$

Next, suppose that $\widehat{A} \neq (\widehat{A})^*$ is a maximal dissipative extension of $\dot{A}$,

$$\mathrm{Im}(\widehat{A}f, f) \geq 0, \quad f \in \mathrm{Dom}(\widehat{A}).$$

Since $\dot{A}$ is symmetric, its dissipative extension $\widehat{A}$ is automatically quasi-selfadjoint [112, 129] (also see [4, 86]), that is,

$$\dot{A} \subset \widehat{A} \subset (\dot{A})^*,$$

and hence,

$$g_+ - \varkappa g_- \in \mathrm{Dom}(\widehat{A}) \quad \text{for some } |\varkappa| < 1. \tag{2.5}$$

By definition, we call $\varkappa$ *the von Neumann parameter of the triple* $(\dot{A}, \widehat{A}, A)$.

Remark 2.1. Likewise, one can think of the von Neumann parameter $\varkappa$ being determined by the dissipative operator $\widehat{A}$ and the pair $\{g_+, g_-\}$ of normalized deficiency elements $g_\pm \in \mathrm{Ker}((\dot{A})^* \mp iI)$. Indeed, given $\widehat{A}$ and $\{g_+, g_-\}$ there are a unique $\varkappa$ satisfying (2.5) and a unique self-adjoint reference extension A of $\dot{A}$ such that (2.1) holds. Therefore, the triple $(\dot{A}, \widehat{A}, A)$ is uniquely determined by the knowledge of $\widehat{A}$ and $\{g_+, g_-\}$. Cleary, $\varkappa$ coincides with the von Neumann parameter of the triple $\varkappa_{(\dot{A}, \widehat{A}, A)}$ which proves the claim.

Given (2.1) and (2.5), consider *the characteristic function* $S(z) = S_{(\dot{A}, \widehat{A}, A)}(z)$ associated with the triple $(\dot{A}, \widehat{A}, A)$ (see [86], cf. [77])

$$S(z) = \frac{s(z) - \varkappa}{\overline{\varkappa}\, s(z) - 1}, \quad z \in \mathbb{C}_+, \tag{2.6}$$

where $s(z) = s_{(\dot{A}, A)}(z)$ is the Livšic function associated with the pair $(\dot{A}, A)$.

By (2.4) and (2.6), one also gets the representation for the characteristic function via the Weyl-Titchmarch function as

$$S(z) = -\frac{1 - \varkappa}{1 - \overline{\varkappa}} \cdot \frac{M(z) - i\frac{1+\varkappa}{1-\varkappa}}{M(z) + i\frac{1+\overline{\varkappa}}{1-\overline{\varkappa}}}. \tag{2.7}$$

We remark that given a triple $(\dot{A}, \widehat{A}, A)$, one can always find a basis $g_\pm$ in the subspace $\mathrm{Ker}(\dot{A}^* - iI) \dotplus \mathrm{Ker}(\dot{A}^* + iI)$ such that

$$\|g_\pm\| = 1, \quad g_\pm \in \mathrm{Ker}((\dot{A})^* \mp iI),$$

$$g_+ - g_- \in \mathrm{Dom}(A)$$

and

$$g_+ - \varkappa g_- \in \mathrm{Dom}(\widehat{A}) \quad \text{for some } |\varkappa| < 1.$$

In this case, the von Neumann parameter $\varkappa$ can explicitly be evaluated in terms of the characteristic function of the triple $(\dot{A}, \widehat{A}, A)$ as

$$\varkappa = S_{(\dot{A}, \widehat{A}, A)}(i). \tag{2.8}$$

Hence, as it follows from (2.6), the Livšic function associated with the pair $(\dot{A}, A)$ admits the representation

$$s_{(\dot{A}, A)}(z) = \frac{S(z) - \varkappa}{\overline{\varkappa} S(z) - 1}, \quad z \in \mathbb{C}_+. \tag{2.9}$$

In particular,

$$S_{(\dot{A}, \widehat{A}, A)}(z) = -s_{(\dot{A}, A)}(z), \quad z \in \mathbb{C}_+,$$

whenever the von Neumann parameter $\varkappa$ of the triple $(\dot{A}, \widehat{A}, A)$ vanishes.

The following proposition provides a curious characterization for the projection of the deficiency subspace $\mathrm{Ker}((\dot{A})^* - zI)$ onto the subspace $\mathrm{Ker}((\dot{A})^* - \overline{z}I)$ along $\mathrm{Dom}(\widehat{A})$ in terms of the characteristic function of the triple.

Proposition 2.2. *Let* $(\dot{A}, \widehat{A}, A)$ *be a triple. Suppose that the deficiency elements* $g_z \in \mathrm{Ker}((\dot{A})^* - zI)$ *and* $g_{\overline{z}} \in \mathrm{Ker}((\dot{A})^* - \overline{z}I)$, $\|g_z\| = \|g_{\overline{z}}\| \neq 0$ *are chosen in such a way that*

$$g_z - \gamma(z)g_{\overline{z}} \in \mathrm{Dom}(\widehat{A}), \quad z \in \mathbb{C}_+. \tag{2.10}$$

Then

$$|\gamma(z)| = |S(z)|, \quad z \in \mathbb{C}_+,$$

where $S(z)$ *is the characteristic function of the triple* $(\dot{A}, \widehat{A}, A)$.

Proof. By Theorem C.1 in Appendix C, the triple $(\dot{A}, \widehat{A}, A)$ is mutually unitarily equivalent to the model triple $(\dot{\mathcal{B}}, \widehat{\mathcal{B}}, \mathcal{B})$ in the Hilbert space $L^2(\mathbb{R}; d\mu)$ given by (C.6), (C.7) and (C.8) in Appendix C. Here $\mu(d\lambda)$ is the measure from the representations

$$M(z) = \int_{\mathbb{R}} \left(\frac{1}{\lambda - z} - \frac{\lambda}{\lambda^2 + 1} \right) d\mu(\lambda), \quad z \in \mathbb{C}_+, \tag{2.11}$$

for the Weyl-Titchmarsh function associated with the pair $(\dot{A}, A)$.

In view of Remark C.2 in Appendix C it suffices to show that if

$$\frac{1}{\lambda - z} - \alpha \frac{1}{\lambda - \overline{z}} \in \mathrm{Dom}(\widehat{\mathcal{B}}), \tag{2.12}$$

then

$$\alpha = |S(z)|, \quad z \in \mathbb{C}_+. \tag{2.13}$$

We claim that

$$\frac{1}{\lambda - z} = \frac{\lambda}{\lambda^2 + 1} + \frac{M(z)}{\lambda^2 + 1} + f(\lambda),$$

where $f \in \mathrm{Dom}(\dot{\mathcal{B}})$.

Indeed,

$$\int_{\mathbb{R}} f(\lambda) d\mu(\lambda) = \int_{\mathbb{R}} \left(\frac{1}{\lambda - z} - \frac{\lambda}{\lambda^2 + 1} \right) d\mu(\lambda) - M(z) \int_{\mathbb{R}} \frac{d\mu(\lambda)}{\lambda^2 + 1}$$

$$= M(z) - M(z) = 0, \quad z \in \mathbb{C}_+.$$

Here we have used (2.11) and the normalization condition (C.5) in Appendix C. From the characterization of the domain of the symmetric operator $\dot{\mathcal{B}}$ given by (C.7) in Appendix C, it follows that $f \in \mathrm{Dom}(\dot{\mathcal{B}})$, proving the claim.

Next, we have

$$\frac{1}{\lambda - z} - \alpha \frac{1}{\lambda - \overline{z}} = \frac{(1 - \alpha)\lambda}{\lambda^2 + 1} + \frac{M(z) - \alpha M(\overline{z})}{\lambda^2 + 1} + h(\lambda), \tag{2.14}$$

where $h \in \mathrm{Dom}(\dot{\mathcal{B}})$.

Recall that by (C.8) in Appendix C,

$$\frac{1}{\lambda - i} - \varkappa \frac{1}{\lambda + i} \in \mathrm{Dom}(\widehat{\mathcal{B}}),$$

where $\varkappa$ is the von Neumann parameter of the triple $(\dot{\mathcal{B}}, \widehat{\mathcal{B}}, \mathcal{B})$. Since the triples $(\dot{A}, \widehat{A}, A)$ and $(\dot{\mathcal{B}}, \widehat{\mathcal{B}}, \mathcal{B})$ are mutually unitarily equivalent, $\varkappa$ coincides with von Neumann parameter of the triple $(\dot{A}, \widehat{A}, A)$ as well.

Since

$$\frac{(1-\alpha)\lambda}{\lambda^2+1} + \frac{M(z)-\alpha M(\bar{z})}{\lambda^2+1} = \frac{1}{2}\left[(1-\alpha)-i(M(z)-\alpha M(\bar{z}))\right]\frac{1}{\lambda-i}$$
$$+ \frac{1}{2}\left[(1-\alpha)-i(M(z)-\alpha M(\bar{z}))\right]\frac{1}{\lambda+i},$$

in view of (2.14), the requirement (2.12) connecting the characteristic functions $S(z)$ of the triple and the Weyl-Tichmarsch function yields the relation

$$-\varkappa = \frac{(1-\alpha)+i(M(z)-\alpha M(\bar{z}))}{(1-\alpha)-i(M(z)-\alpha M(\bar{z}))}, \quad z \in \mathbb{C}_+.$$

Hence,

$$\alpha = \frac{F(z)}{F(\bar{z})},$$

where

$$F(z) = 1 + \varkappa + iM(z)(1-\varkappa).$$

On the other hand, from the relation (2.7) it follows that

$$S(z) = \frac{F(z)}{\overline{F(\bar{z})}}$$

and therefore

$$|\alpha| = |S(z)|, \quad z \in \mathbb{C}_+,$$

which proves (2.13) and completes the proof of the proposition. $\qquad\square$

Remark 2.3. Notice that given the deficiency elements $\|g_z\| = \|g_{\bar{z}}\| \neq 0$, $z \in \mathbb{C}_+$, one can always find $\gamma(z)$ such that (2.10) holds. Indeed, from the definition of the quasi-selfadjoint extension $\widehat{A}$ it follows that one can find α and β such that

$$0 \neq \alpha g_z + \beta g_{\bar{z}} \in \text{Dom}(\widehat{A}).$$

Therefore, it suffices to show that $\alpha \neq 0$. Otherwise, $g_{\bar{z}} \in \text{Dom}(\widehat{A})$ and hence

$$\text{Im}(\widehat{A}g_{\bar{z}}, g_{\bar{z}}) = \text{Im}((\dot{A})^* g_{\bar{z}}, g_{\bar{z}}) = \text{Im}(\bar{z}g_{\bar{z}}, g_{\bar{z}}) = \text{Im}(\bar{z})\|g_{\bar{z}}\|^2 < 0,$$

which contradicts the requirement that $\widehat{A}$ is a dissipative operator.

Definition 2.4. We call the harmonic function

$$\Gamma_{\widehat{A}}(z) = \log|\gamma(z)| = \log|S(z)|, \quad z \in \mathbb{C}_+ \setminus \{z_k \mid S(z_k) = 0\}, \qquad (2.15)$$

the *von Neumann* (*logarithmic*) *potential* of the dissipative operator $\widehat{A}$. Here $\gamma(z)$ is the function referred to in Proposition 2.2.

Remark 2.5. We remark that under the assumption that the symmetric operator $\dot{A}$ is prime, both the Weyl-Titchmarsh function $M(z)$ and the Livšic function $s(z)$ are complete unitary invariants of the pair $(\dot{A}, A)$. Moreover, in this case, the characteristic function $S(z)$ is a complete unitary invariant of the triple $(\dot{A}, \widehat{A}, A)$ (see [86], also see Theorem C.1 in Appendix C). Notice that if the symmetric operator $\dot{A}$ from a triple $(\dot{A}, \widehat{A}, A)$ in the Hilbert space $\mathcal{H}$ is not prime and $\dot{A}'$ is its prime part in a subspace $\mathcal{H}' \subset \mathcal{H}$, then the triples $(\dot{A}, \widehat{A}, A)$ and $(\dot{A}|_{\mathcal{H}'}, \widehat{A}|_{\mathcal{H}'}, A|_{\mathcal{H}'})$ have the same characteristic function (see Theorem B.2 in Appendix B for details).

As it follows from Lemma E.1 and Proposition E.2 in Appendix E, the absolute value of the Livšic function $|s(z)|$ is a complete unitary invarinat of the symmetric operator $\dot{A}$ while $|S(z)|$ is a (complete) unitary invariant of the maximal dissipative operator $\widehat{A}$. In particular, in view of (2.8), the von Neumann parameter $\varkappa_{(\dot{A}, \widehat{A}, A)}$ is a unitary invariant of the triple $(\dot{A}, \widehat{A}, A)$, while its absolute value

$$\widehat{\kappa}(\widehat{A}) = |\varkappa_{(\dot{A}, \widehat{A}, A)}| \tag{2.16}$$

is a well defined unitary invariant of the dissipative operator $\widehat{A}$.

We also refer to [73] where it was shown that the knowledge or $s(z)$ and $S(z)$ (up to a unimodular constant factor), equivalently, $|s(z)|$ and $|S(z)|$ characterizes the symmetric operator $\dot{A}$ and is maximal dissipative extension $\widehat{A}$, respectively, up to unitary equivalence (whenever $\dot{A}$ is a prime operator).

We also notice that the knowledge of the von Neumann logarithmic potential $\Gamma_{\widehat{A}}(z)$ determines the dissipative operator $\widehat{A}$ up to unitary equivalence.

Chapter 3

THE COMMUTATION RELATIONS AND CHARACTER-AUTOMORPHIC FUNCTIONS

Throughout this chapter we assume the following hypothesis.

Hypothesis 3.1. Assume that $\dot{A}$ is a symmetric operator with deficiency indices $(1, 1)$ and $\mathbb{R} \ni t \mapsto U_t$ a strongly continuous unitary group. Suppose, in addition, that $\mathrm{Dom}(\dot{A})$ is U_t-invariant and the commutation relations

$$U_t^* \dot{A} U_t = \dot{A} + tI \quad \text{on} \quad \mathrm{Dom}(\dot{A}), \quad t \in \mathbb{R}, \tag{3.1}$$

hold.

It is well known that under Hypothesis 3.1 the symmetric operator $\dot{A}$ has either a) a self-adjoint A or/and b) a non-selfadjoint maximal dissipative extension $\widehat{A}$ satisfying the same commutation relations (cf. [52, Theorem 15], also see [88, Theorem 5.4]).

It is worth mentioning that the existence of a quasi-selfadjoint extension of $\dot{A}$ that solves (3.1) required in Hypothesis 3.1 is a consequence of the Lefschetz fixed point theorem for flows on manifolds:

Proposition 3.2 (see, e.g., [136, Theorem 6.28]). *If $\mathcal{M}$ is a closed oriented manifold such that the Euler characteristic $\chi(\mathcal{M})$ of $\mathcal{M}$ is not zero, then any flow on $\mathcal{M}$ has a fixed point.*

Indeed, if $\dot{A}$ satisfies Hypothesis 3.1 and $\widehat{A}$ is a maximal dissipative extension of $\dot{A}$, then $\widehat{A}_t = U_t \widehat{A} U_t^*$ also extends $\dot{A}$. Since the set of maximal dissipative extensions of $\dot{A}$ is in one-to-one correspondence with the closed unit disk $\overline{\mathbb{D}}$, and

$$\widehat{A} \mapsto \widehat{A}_t \tag{3.2}$$

determines a flow $\varphi(t, \cdot)$ on $\overline{\mathbb{D}}$ (the continuity of the flow can easily be established (see, e.g., [88])). By the Lefschetz theorem, the flow $\varphi(t, \cdot)$ has a fixed point either on the boundary of the unit disk or in its interior.

First, consider the case where the flow $\varphi(t, \cdot)$ has a fixed point on the boundary of the unit disk and therefore the commutation relations (3.1) have a self-adjoint solution.

Theorem 3.3. *Assume Hypothesis* 3.1. *Suppose that A is a self-adjoint extension of the symmetric operator $\dot{A}$ such that*

$$U_t^* A U_t = A + tI \quad on \quad \mathrm{Dom}(A). \tag{3.3}$$

Then the Weyl-Titchmarsh function $M(z)$ of the pair $(\dot{A}, A)$ has the form

$$M(z) = i, \quad z \in \mathbb{C}_+.$$

Equivalently, the Livšic function $s(z)$ associated with the pair $(\dot{A}, A)$ vanishes identically in the upper half-plane,

$$s(z) = 0, \quad z \in \mathbb{C}_+.$$

Proof. Introducing the family of bounded operators

$$B_t = U_t(A - iI)(A - iI + tI)^{-1}, \quad t \in \mathbb{R},$$

it is easy to see that the family $\mathbb{R} \ni t \mapsto B_t$ forms a strongly continuous (commutative) group. Indeed, using the commutation relation (3.3) for the self-adjoint operator A one obtains

$$\begin{aligned}
B_t B_s &= U_t(A - iI)(A + (-i + t)I)^{-1} U_s(A - iI)(A + (-i + s)I)^{-1} \\
&= U_t U_s U_s^*(A - iI)(A - iI + tI)^{-1} U_s(A - iI)(A + (-i + s)I)^{-1} \\
&= U_{t+s}(A - iI + sI)(A - iI + (t + s)I)^{-1}(A - iI)(A - iI + sI)^{-1} \\
&= U_{t+s}(A - iI + (t + s)I)^{-1}(A - iI) \\
&= B_{t+s}, \quad s, t \in \mathbb{R}.
\end{aligned}$$

Let $g_+ \in \mathrm{Ker}((\dot{A})^* - iI)$, $\|g_+\| = 1$, be a normalized deficiency element of $\dot{A}$. Since

$$(A - iI)(A - (i - t)I)^{-1} g_+ \in \mathrm{Ker}((\dot{A})^* - (i - t)I)$$

and

$$U_t \operatorname{Ker}((\dot{A})^* - (i - t)I) = \operatorname{Ker}((\dot{A})^* - iI),$$

one concludes that the deficiency subspace $\operatorname{Ker}((\dot{A})^* - iI)$ is invariant for B_t, $t \in \mathbb{R}$. Therefore, the restriction B_t on the deficiency subspace is a continuous one-dimensional representation (a one-dimensional representation of a strongly continuous group is continuous). Hence,

$$B_t g_+ = b^t g_+ \quad \text{for some} \quad b \in \mathbb{C}.$$

From the definition of the Weyl-Titchmarsh function (2.2) it follows that

$$M'(z) := \frac{d}{dz} M(z) = ((A^2 + I)(A - zI)^{-2} g_+, g_+), \quad z \in \mathbb{C}_+.$$

One computes

$$
\begin{aligned}
|b|^{2t} M'(i - t) &= |b|^{2t}((A^2 + I)(A - iI + tI)^{-2} g_+, g_+) \\
&= ((A^2 + I)(A - iI + tI)^{-2} B_t g_+, B_t g_+) \\
&= (B_t^*(A^2 + I)(A - iI + tI)^{-2} B_t g_+, g_+) \\
&= ((A + iI)(A + iI + tI)^{-1} U_t^*(A^2 + I) \\
&\quad \times (A - iI - tI)^{-2} B_t g_+, g_+) \\
&= ((A + iI)(A + iI + tI)^{-1} U_t^*(A^2 + I)(A - iI - tI)^{-2} U_t \\
&\quad \times (A - iI)(A - iI + tI)^{-1} g_+, g_+) \\
&= ((A + iI)(A + iI + tI)^{-1}((A + tI)^2 + I)(A - iI)^{-2}(A - iI) \\
&\quad \times (A - iI + tI)^{-1} g_+, g_+) \\
&= ((A + iI)(A - iI)^{-1} g_+, g_+) \\
&= (g_+, (A - iI)(A + iI)^{-1} g_+).
\end{aligned}
$$

Hence,

$$M'(i - t) = |b|^{-2t}(g_+, g_-), \quad t \in \mathbb{R}, \tag{3.4}$$

where

$$g_- = (A - iI)(A + iI)^{-1} g_+ \in \operatorname{Ker}((\dot{A})^* + iI).$$

Denote by $\mu(d\lambda)$ the spectral measure of the element g_+, that is,

$$\mu(d\lambda) = (E_A(d\lambda) g_+, g_+),$$

where $E_A(d\lambda)$ is the projection-valued spectral measure of the self-adjoint operator A from the spectral decomposition

$$A = \int_{\mathbb{R}} \lambda dE_A(\lambda).$$

Since

$$M(z) = \int_{\mathbb{R}} \frac{\lambda z + 1}{\lambda - z} d\mu(\lambda),$$

and therefore

$$M'(i - t) = \int_{\mathbb{R}} \frac{\lambda^2 + 1}{(\lambda - t - i)^2} d\mu(\lambda),$$

one gets the estimate

$$|M'(i - t)| \leq \int_{\mathbb{R}} \frac{\lambda^2 + 1}{(\lambda + t)^2 + 1} d\mu(\lambda)$$

$$= \int_{\{|\lambda| \leq 2|t|\}} \frac{\lambda^2 + 1}{(\lambda + t)^2 + 1} d\mu(\lambda) + \int_{\{|\lambda| > 2|t|\}} \frac{\lambda^2 + 1}{(\lambda + t)^2 + 1} d\mu(\lambda)$$

$$\leq (4t^2 + 1)\mu\{|\lambda| \leq |t|\} + \left(\frac{1}{(1 - \frac{1}{2})^2} + 1 \right) \mu\{|\lambda| > |t|\}.$$

Therefore,

$$M'(i - t) = O(t^2) \quad \text{as} \quad t \to \infty. \tag{3.5}$$

Combining (3.5) with (3.4) shows that either $(g_+, g_-) = 0$ or $|b| = 1$.

In the first case, i.e. $(g_+, g_-) = 0$, $M(z)$ is a constant function and hence

$$M(z) = M(i) = i, \quad z \in \mathbb{C}.$$

If $|b| = 1$, we have

$$M'(i - t) = (g_+, g_-), \quad t \in \mathbb{R}.$$

In particular, $M'(z) = (g_+, g_-)$ for all $z \in \mathbb{C}_+$ and hence

$$M(z) = (g_+, g_-)z + C, \quad z \in \mathbb{C}_+,$$

for some constant C. We have, see [56],

$$\lim_{y \to \infty} \frac{M(iy)}{y} = 0,$$

which implies $(g_+, g_-) = 0$ and again shows that $M(z)$ is a constant function in the upper half-plane and

$$M(z) = C = M(i) = i, \quad z \in \mathbb{C}_+,$$

which completes the proof. $\square$

Remark 3.4. In connection with our main hypothesis of this chapter, we remark that if a self-adjoint operator A solves commutation relations (3.3), then one can always find a symmetric restriction $\dot{A}$ that satisfies Hypothesis 3.1. Indeed, the commutation relations (3.3) imply that the one-parameter group $V_s = e^{isA}$ generated by A satisfies commutation relations in the Weyl form (see, e.g., [27, Ch. 3, Sect. 1, Theorem 5] or [125]) and then the existence of such a symmetric operator $\dot{A}$ is an immediate corollary of the Stone-von Neumann uniqueness theorem.

Next, we threat the case where the flow $\varphi(t, \cdot)$ associated with the transformation (3.2) has a fixed point in the interior of the unit disk. That is, $\dot{A}$ admits a maximal dissipative "invariant" extension that is not self-adjoint (cf. [55, Theorem 20]). We present the corresponding result in a slightly stronger form. In particular, in this case the requirement (3.1) can be relaxed.

Theorem 3.5. *Suppose that $\widehat{A}$ is a quasi-selfadjoint dissipative extension of a closed symmetric operator $\dot{A}$ with deficiency indices $(1, 1)$ and A is a (reference) self-adjoint extension of $\dot{A}$.*

Suppose that the commutation relation

$$U_t^* \widehat{A} U_t = \widehat{A} + tI \quad on \quad \mathrm{Dom}(\widehat{A}) \tag{3.6}$$

hold.

Then the characteristic function $S(z)$ associated with the triple $(\dot{A}, \widehat{A}, A)$ admits the representation

$$S(z) = ke^{i\ell z}, \quad z \in \mathbb{C}_+, \tag{3.7}$$

for some $|k| \leq 1$ and $\ell \geq 0$. Furthermore, if $\ell = 0$, then necessarily $|k| < 1$ and if $|k| = 1$, then $\ell > 0$.

In particular, the von Neumann parameter $\varkappa$ of the triple $(\dot{A}, \widehat{A}, A)$ is given by

$$\varkappa = ke^{-\ell}.$$

Proof. Since $\widehat{A}$ is a quasi-selfadjoint extension of $\dot{A}$, we have

$$\dot{A} = \widehat{A}\big|_{\mathrm{Dom}(\widehat{A}) \cap \mathrm{Dom}((\widehat{A})^*)}. \tag{3.8}$$

We claim that

$$U_t^* (\widehat{A})^* U_t = (\widehat{A})^* + tI \quad on \quad \mathrm{Dom}((\widehat{A})^*). \tag{3.9}$$

To see that assume that $g \in \mathrm{Dom}(\widehat{A})$ and $f \in \mathrm{Dom}((\widehat{A})^*)$. Then

$$(\widehat{A}g, U_t f) = (U_t^* \widehat{A}g, f) = (U_t^* \widehat{A} U_t U_t^* g, f) = ((\widehat{A} + tI) U_t^* g, f)$$
$$= (U_t^* g, ((\widehat{A})^* + tI)f) = (g, U_t((\widehat{A})^* + tI)f). \qquad (3.10)$$

Here we have used that the domain $\mathrm{Dom}(\widehat{A})$ is U_t-invariant for all $t \in \mathbb{R}$ and therefore $U_t^* g \in \mathrm{Dom}(\widehat{A})$. Since (3.10) holds for all $g \in \mathrm{Dom}(\widehat{A})$, one ensures that $\mathrm{Dom}((\widehat{A})^*)$ is U_t-invariant and

$$(\widehat{A})^* U_t f = U_t((\widehat{A})^* + tI)f, \quad f \in \mathrm{Dom}((\widehat{A})^*),$$

which proves (3.9).

Since $U_t(\mathrm{Dom}((\widehat{A})^*)) = \mathrm{Dom}((\widehat{A})^*)$, from (3.8) one concludes that the commutation relations

$$U_t^* (\dot{A})^* U_t = (\dot{A})^* + tI \quad \text{on} \quad \mathrm{Dom}((\dot{A})^*) \qquad (3.11)$$

hold.

Taking into account that

$$\widehat{A} = U_t(\widehat{A} + tI)U_t^*$$

and

$$\dot{A} = U_t(\dot{A} + tI)U_t^*,$$

and also observing that the operator A_t given by

$$A_t = U_t(A + tI)U_t^*$$

is a self-adjoint extension of the symmetric operator $\dot{A}$, we see that the triples $(\dot{A}, \widehat{A}, A_t)$ and $(\dot{A} + tI, \widehat{A} + tI, A + tI)$ are mutually unitarily equivalent. In particular,

$$S_{(\dot{A}+tI, \widehat{A}+tI, A+tI)}(z) = S_{(\dot{A}, \widehat{A}, A_t)}(z), \quad z \in \mathbb{C}_+.$$

Since A_t is a self-adjoint extension of $\dot{A}$, by Lemma E.1 (see (E.4)) in Appendix E, we have

$$S_{(\dot{A}, \widehat{A}, A_t)}(z) = \Theta_t^{(1)} S_{(\dot{A}, \widehat{A}, A)}(z), \quad z \in \mathbb{C}_+,$$

for some unimodular constant $\Theta_t^{(1)}$, $|\Theta_t^{(1)}| = 1$, which is a continuous function of the parameter t (see [88] for the proof of continuity).

By Theorem F.1 in Appendix F,

$$S_{(\dot{A}+tI,\widehat{A}+tI,A+tI)}(z) = \Theta_t^{(2)} S_{(\dot{A},\widehat{A},A)}(z-t), \quad z \in \mathbb{C}_+,$$

where $\Theta_t^{(2)}$ is another continuous unimodular function in t. Therefore, the functional equation

$$S_{(\dot{A},\widehat{A},A)}(z-t) = \Theta_t S_{(\dot{A},\widehat{A},A)}(z), \quad t \in \mathbb{R},$$

holds, where

$$\Theta_t = \Theta_t^{(1)} \overline{\Theta_t^{(2)}}.$$

From the functional equation it also follows that Θ_t is a continuous unimodular solution of the equation

$$\Theta_{t+s} = \Theta_t \Theta_s, \quad s, t \in \mathbb{R},$$

and therefore (see, e.g., [32, XVII, 6])

$$\Theta_t = e^{i\ell t} \quad \text{for some} \quad \ell \in \mathbb{R}.$$

In particular, this proves that the characteristic function $S_{(\dot{A},\widehat{A},A)}(z)$ is a character-automorphic function with respect to the shifts, that is

$$S_{(\dot{A},\widehat{A},A)}(z+t) = e^{i\ell t} S_{(\dot{A},\widehat{A},A)}(z), \quad t \in \mathbb{R}. \tag{3.12}$$

Since $S_{(\dot{A},\widehat{A},A)}(z)$ is a contractive analytic function on $\mathbb{C}_+$, it admits the representation

$$S_{(\dot{A},\widehat{A},A)}(z) = \theta B(z) e^{iM(z)},$$

where $|\theta| \leq 1$, B is the Blaschke product associated with the (possible) zeros of $S_{(\dot{A},\widehat{A},A)}(z)$ in the upper half-plane, and $M(z)$ is a Herglotz-Nevanlinna function.

Suppose that the characteristic function $S_{(\dot{A},\widehat{A},A)}(z)$ is not identically zero and thus $\theta \neq 0$. Then, from the functional equation (3.12) it follows that $S_{(\dot{A},\widehat{A},A)}(z)$ has no zeros in $\mathbb{C}_+$, and hence

$$S_{(\dot{A},\widehat{A},A)}(z) = \theta e^{iM(z)}, \quad z \in \mathbb{C}_+.$$

Since $S_{(\dot{A},\widehat{A},A)}(z)$ is character-automorphic, one concludes that the functional equation

$$M(z+t) = \ell t + M(z)$$

holds.

Next, we have

$$\frac{d}{dt}M(z+t)|_{t=0} = M'(z) = \ell.$$

Taking into account that $M(z)$ maps the upper half-plane into itself, we obtain that

$$M(z) = \ell z + b,$$

where $\ell \geq 0$ and $\mathrm{Im}(b) \geq 0$. Therefore,

$$S_{(\dot{A},\widehat{A},A)}(z) = \theta e^{ib} e^{i\ell z},$$

which proves (3.7) with $k = \theta e^{ib}$.   □

Remark 3.6. Notice that in the situation of Theorem 3.5 we have that

$$\dot{A} = \widehat{A}|_{\mathrm{Dom}(\widehat{A}) \cap \mathrm{Dom}(\widehat{A}^*)}.$$

Therefore the symmetric operator $\dot{A}$ is uniquely determined by the generator $\widehat{A}$. In this case we will call $\dot{A}$ the symmetric part of $\widehat{A}$.

We also remark that if under the hypothesis of Theorem 3.5 there is no self-adjoint extension satisfying the commutation relations (3.3), the corresponding maximal dissipative extension $\widehat{A}$ is unique (see, e.g., [88, Theorem 6.3]).

The following corollary is the first step towards a complete classification up to unitary equivalence of "invariant" symmetric operators from Hypothesis 3.1 (see Problem (I) a) in Introduction).

Corollary 3.7. *Assume Hypothesis* 3.1. *Then there exists a self-adjoint extension A of $\dot{A}$ such that the Livšic function associated with the pair $(\dot{A}, A)$ admits the representation*

$$s_{(\dot{A},A)}(z) = k \frac{e^{iz\ell} - e^{-\ell}}{k^2 e^{-\ell} e^{iz\ell} - 1}, \quad z \in \mathbb{C}_+,$$

for some $0 \leq k \leq 1$ and $\ell > 0$.

Proof. If $\dot{A}$ admits a self-adjoint extension A that satisfies the same commutation relations as $\dot{A}$ does, then by Theorem 3.3

$$s_{(\dot{A},A)}(z) = 0, \quad z \in \mathbb{C}_+,$$

which proves the claimed representation with $k = 0$.

If $\dot{A}$ admits a maximal (non-selfadjoint) dissipative extension $\widehat{A}$ that satisfies the same commutation relations, then by Theorem 3.5 there exists a (reference) self-adjoint extension A' of $\dot{A}$ such that the characteristic function of the triple $(\dot{A}, \widehat{A}, A')$ is of the form

$$S(z) = k e^{i\ell z}, \quad z \in \mathbb{C}_+.$$

Therefore, by (2.9)

$$s_{(\dot{A}, A')}(z) = \frac{S(z) - S(i)}{\overline{S(i)}\, S(z) - 1} = k \frac{e^{iz\ell} - e^{-\ell}}{|k|^2 e^{-\ell} e^{iz\ell} - 1}.$$

By Lemma E.1 in Appendix E, one can always find a possibly different self-adjoint extension A of $\dot{A}$ such that

$$s_{(\dot{A}, A)}(z) = |k| \frac{e^{iz\ell} - e^{-\ell}}{|k|^2 e^{-\ell} e^{iz\ell} - 1}, \quad z \in \mathbb{C}_+,$$

and the claim follows. $\qquad\square$

Chapter 4

THE DIFFERENTIATION OPERATOR ON METRIC GRAPHS

Let $\mathbb{Y}$ be a directed metric graph (see, e.g., [8, 59]). We will distinguish the following three cases.

Case (i):

$$\mathbb{Y} = (-\infty, 0) \sqcup (0, \infty),$$

with $(-\infty, 0)$ the incoming and $(0, \infty)$ outgoing bonds;

Case (ii):

$$\mathbb{Y} = (0, \ell), \quad \text{the outgoing bond;}$$

Case (iii):

$$\mathbb{Y} = (-\infty, 0) \sqcup (0, \infty) \sqcup (0, \ell),$$

with $(-\infty, 0)$ the incoming and both $(0, \infty)$ and $(0, \ell)$ the outgoing bonds.

Denote by $\dot{D} = i\frac{d}{dx}$ the differentiation operator on the metric graph $\mathbb{Y}$ in Cases (i)–(iii) defined on the domain $\mathrm{Dom}(\dot{D})$ of functions $f \in W_2^1(\mathbb{Y})$ with the following boundary conditions on the vertices of the graph, in Case (i):

$$f_\infty(0+) = f_\infty(0-) = 0; \tag{4.1}$$

in Case (ii):

$$f_\ell(0) = f_\ell(\ell) = 0; \tag{4.2}$$

in Case (iii):

$$\begin{cases} f_\infty(0+) & = k f_\infty(0-) \\ f_\ell(0) & = \sqrt{1 - k^2} f_\infty(0-) \quad \text{for some} \quad 0 < k < 1. \\ f_\ell(\ell) & = 0 \end{cases} \tag{4.3}$$

Here we have used the following notation.

If the graph Υ is in Cases (i) and (ii), the functions from the Hilbert space $L^2(\Upsilon)$ are denoted by f_∞ and f_ℓ, respectively.

If the metric graph Υ is in Case (iii), in view of the natural identification of $L^2(\Upsilon)$ with the orthogonal sum $L^2(\mathbb{R}) \oplus L^2((0, \ell))$, it is convenient to represent an arbitrary element $f \in L^2(\Upsilon)$ as the two-component vector-function

$$f = \begin{pmatrix} f_\infty \\ f_\ell \end{pmatrix}.$$

(Here $L^2(\Upsilon)$ denotes the Hilbert space of square-integrable functions with respect to Lebesgue measure on the edges of the metric graph Υ.)

Notice that if the graph Υ is in Case (iii) and $k = 0$ in (4.3), then the boundary conditions (4.3) can be rewritten as

$$\begin{cases} f_\infty(0+) = 0 \\ f_\ell(0) = f_\infty(0-) \\ f_\ell(\ell) = 0. \end{cases}$$

In this case, the operator $\dot{D}$ splits into the orthogonal sum of the symmetric differentiation operators on the semi-axes $(-\infty, \ell)$ and $(0, \infty)$ with the Dirichlet boundary conditions at the end-points, respectively. Therefore, if $k = 0$, then the operator $\dot{D}$ is unitarily equivalent to the symmetric differentiation in Case (i).

Remark 4.1. In Cases (i) and (iii) the metric graph Υ is not finite. However, one can assign two additional vertices to the external edges at $\pm\infty$. Under this hypothesis, in all Cases (i)–(iii) the Euler characteristic $\chi(\Upsilon)$ of the graph Υ, the number of vertices minus the number of edges, equals one. Therefore, the corresponding first Betti number $\beta(\Upsilon) = -\chi(\Upsilon) + 1$ of the graph Υ, the number of edges that have to be removed to turn the graph into a connected tree, vanishes.

Lemma 4.2. *The operator $\dot{D}$ on a metric graph $\mathbb{Y}$ is Cases (i)–(iii) is symmetric. Moreover,*

$$\text{Dom}((\dot{D})^*) = \begin{cases} W_2^1((-\infty,0)) \oplus W_2^1((0,\infty)), & \text{in Case } (i) \\ W_2^1((0,\ell)), & \text{in Case } (ii) \end{cases}.$$

In Case (iii), the domain $\text{Dom}((\dot{D})^*)$ consists of the vector-functions

$$h = (h_\infty, h_\ell)^T \in (W_2^1(\mathbb{R}_-) \oplus W_2^1(\mathbb{R}_+)) \oplus W_2^1((0,\ell))$$

satisfying the "boundary condition"

$$h_\infty(0-) - k\, h_\infty(0+) - \sqrt{1-k^2}\, h_\ell(0) = 0. \tag{4.4}$$

Proof. The corresponding result in Cases (i) and (ii) is well known.

In Case (iii), from (4.3) it follows that for $f = (f_\infty, f_\ell)^T \in \text{Dom}(\dot{D})$ the "quantum Kirchhoff rule"

$$|f_\infty(0-)|^2 = |f_\infty(0+)|^2 + |f_\ell(0)|^2$$

holds. Since also $f_\ell(\ell) = 0$, integration by parts

$$\begin{aligned}
(\dot{D}f, f) &= \int_{-\infty}^{0} i f'_\infty(x)\overline{f_\infty(x)}dx + \int_{0}^{\infty} i f'_\infty(x)\overline{f_\infty(x)}dx + \int_{0}^{\ell} i f'_\ell(x)\overline{f_\ell(x)}dx \\
&= -\int_{-\infty}^{0} i f_\infty(x)\overline{f'_\infty(x)}dx - \int_{0}^{\infty} i f_\infty(x)\overline{f'_\infty(x)}dx \\
&\quad - \int_{0}^{\ell} i f_\ell(x)\overline{f'_\ell(x)}dx \\
&\quad + i\left(|f_\infty(0-)|^2 - |f_\infty(0+)|^2 - |f_\ell(0)|^2\right), \quad f \in \text{Dom}(\dot{D}),
\end{aligned}$$

shows that the quadratic form $(\dot{D}f, f)$ is real and therefore the operator $\dot{D}$ is indeed symmetric.

Similar computations show that

$$\begin{aligned}
(h, \dot{D}f) &= \int_{-\infty}^{0} i h'_\infty(x)\overline{f_-(x)}dx + \int_{0}^{\infty} i h'_\infty(x)\overline{f_+(x)}dx + \int_{0}^{\ell} i h'_\ell(x)\overline{f_\ell(x)}dx \\
&= -\int_{-\infty}^{0} i h_\infty(x)\overline{f'_\infty(x)}dx - \int_{0}^{\infty} i h_\infty(x)\overline{f'_\infty(x)}dx \\
&\quad - \int_{0}^{\ell} i h_\ell(x)\overline{f'_\ell(x)}dx + i\left(h_\infty(0-)\overline{f_\infty(0-)}\right. \\
&\quad \left. - h_\infty(0+)\overline{f_\infty(0+)} - h_\ell(0)\overline{f_\ell(0)}\right), \quad f \in \text{Dom}(\dot{D}).
\end{aligned}$$

Therefore, $h \in \mathrm{Dom}((\dot{D})^*)$ if and only if

$$h_\infty(0-)\overline{f_\infty(0-)} - h_\infty(0+)\overline{f_\infty(0+)} - h_\ell(0)\overline{f_\ell(0)} = 0 \quad \text{for all } f \in \mathrm{Dom}(\dot{D}).$$

Taking into account the boundary conditions (4.3), we have

$$\left(h_\infty(0-) - h_\infty(0+)k - h_\ell(0)\sqrt{1-k^2}\right)\overline{f_\infty(0-)} = 0 \tag{4.5}$$

for all $f \in \mathrm{Dom}(\dot{D})$. Since $f_\infty(0-)$ may be chosen arbitrarily, (4.4) follows from (4.5). $\quad\square$

The following lemma introduces a natural (standard) basis in the subspace

$$\mathcal{N} = \mathrm{Ker}((\dot{D})^* - iI) \dot{+} \mathrm{Ker}((\dot{D})^* + iI).$$

Lemma 4.3. *The deficiency subspaces* $\mathrm{Ker}((\dot{D})^* \mp iI)$ *of the symmetric operator* $\dot{D}$ *on the metric graph* $\mathbb{Y}$ *is Cases (i)–(iii) are spanned by the following normalized deficiency elements* $g_\pm$. *Here,*
in Case (i),

$$g_+(x) = \sqrt{2}e^x \chi_{(-\infty,0)}(x) \quad and \quad g_-(x) = \sqrt{2}e^{-x}\chi_{(0,\infty)}(x), \quad x \in \mathbb{R}, \tag{4.6}$$

in Case (ii),

$$g_+(x) = \frac{\sqrt{2}}{\sqrt{e^{2\ell}-1}}e^x \quad and$$
$$g_-(x) = \frac{\sqrt{2}}{\sqrt{e^{2\ell}-1}}e^{\ell-x}, \quad x \in [0,\ell]. \tag{4.7}$$

Finally, in Case (iii),

$$g_+(x) = \frac{\sqrt{2}}{\sqrt{e^{2\ell}-k^2}}e^x \begin{cases} \sqrt{1-k^2}\chi_{(-\infty,0)}(x), & x \in (-\infty,0) \sqcup (0,\infty) \\ 1, & x \in [0,\ell] \end{cases}, \tag{4.8}$$

$$g_-(x) = \frac{\sqrt{2}}{\sqrt{e^{2\ell}-k^2}}e^{\ell-x} \begin{cases} -\sqrt{1-k^2}\chi_{(0,\infty)}(x), & x \in (-\infty,0) \sqcup (0,\infty) \\ k, & x \in [0,\ell] \end{cases}. \tag{4.9}$$

In particular, $\dot{D}$ *is a symmetric operator with deficiency indices* $(1,1)$.

Proof. The deficiency subspaces of the symmetric operator $\dot{D}$ in Cases (i) and (ii) can be easily calculated.

Indeed, in Case (i), we have

$$\mathrm{Ker}((\dot{D})^* - zI) = \mathrm{span}\{g_z\},$$

where

$$g_z(x) = \begin{cases} e^{-izx}, & x < 0 \\ 0, & x \geq 0 \end{cases} \quad \text{and} \quad g_z(x) = \begin{cases} 0, & x < 0 \\ e^{-izx}, & x \geq 0 \end{cases} \quad (4.10)$$

for $z \in \mathbb{C}_+$ and $z \in \mathbb{C}_-$, respectively, which proves (4.6).

In Case (ii),

$$\mathrm{Ker}((\dot{D})^* - zI) = \mathrm{span}\{g_z\},$$

where

$$g_z(x) = e^{-izx}, \quad x \in [0, \ell], \quad z \in \mathbb{C} \setminus \mathbb{R}, \quad (4.11)$$

and (4.7) follows.

In Case (iii), from the description of $\mathrm{Dom}((\dot{D})^*)$ provided by Lemma 4.2 it follows that the deficiency subspace $\mathrm{Ker}((\dot{D})^* - zI) = \mathrm{span}\{g_z\}$, $z \in \mathbb{C} \setminus \mathbb{R}$, is generated by the functions

$$g_z(x) = \begin{cases} \sqrt{1 - k^2}e^{-izx}\chi_{(-\infty,0)}(x), & x \in (-\infty, 0) \sqcup (0, \infty) \\ e^{-izx}, & x \in [0, \ell] \end{cases}, \quad z \in \mathbb{C}_+,$$

$$(4.12)$$

and

$$g_z(x) = \begin{cases} -\sqrt{1 - k^2}e^{-izx}\chi_{(0,\infty)}(x), & x \in (-\infty, 0) \sqcup (0, \infty) \\ ke^{-izx}, & x \in [0, \ell] \end{cases}, \quad z \in \mathbb{C}_-,$$

$$(4.13)$$

proving (4.8) and (4.9). □

Recall (see Appendix B) that a symmetric operator $\dot{A}$ is called a prime operator if there is no (non-trivial) subspace invariant under $\dot{A}$ such that the restriction of $\dot{A}$ to this subspace is self-adjoint.

Lemma 4.4. *The symmetric differentiation operator $\dot{D}$ on the metric graph $\mathbb{Y}$ in Cases (i)–(iii) is a prime operator.*

Proof. In Cases (i) and (ii) the corresponding result is known (see [3]).
Suppose therefore that $\dot{D}$ is in Case (iii).

First, we show that if $f \in L^2(\mathbb{Y}) = L^2(\mathbb{R}) \oplus L^2((0, \ell))$ and

$$(f, g_z) = 0, \quad \text{for all } z \in \mathbb{C} \setminus \mathbb{R}, \tag{4.14}$$

then necessarily $f = 0$.

Indeed, suppose that $f = (f_\infty, f_\ell)^T$, with $f_\infty \in L^2(\mathbb{R})$ and $f_\ell \in L^2((0, \ell))$ and let (4.14) hold. In particular, for all $0 \neq s \in \mathbb{R}$

$$(f, g_{is}) = 0.$$

Then, given the description of the deficiency subspaces (4.12) and (4.13), one gets that

$$\sqrt{1 - k^2} \int_{-\infty}^0 f_\infty(x) e^{sx} dx + \int_0^\ell f_\ell(x) e^{sx} dx = 0 \quad \text{for all } s > 0, \tag{4.15}$$

and

$$-\sqrt{1 - k^2} \int_0^\infty f_\infty(x) e^{-sx} dx + k \int_0^\ell f_\ell(x) e^{-sx} dx = 0 \quad \text{for all } s > 0. \tag{4.16}$$

Therefore, from (4.15) it follows that

$$\int_{-\infty}^\ell h(x) e^{sx} dx = 0 \quad \text{for all } s > 0,$$

where

$$h(x) = \begin{cases} \sqrt{1 - k^2} f_\infty(x), & x < 0 \\ f_\ell(x), & x \in [0, \ell] \end{cases}.$$

By the uniqueness theorem for the Laplace transformation (see, e.g., [22, Theorem 5.1]), we have that $h(x) = 0$ almost everywhere $x \in \mathbb{R}$. Since $k \neq 1$ (see (4.3)), we have

$$f_\infty(x) = 0, \quad \text{a.e. } x \in (-\infty, 0)$$

and

$$f_\ell(x) = 0, \quad \text{a.e. } x \in [0, \ell].$$

Then, from (4.16) it follows that

$$\int_0^\infty f_\infty(x) e^{-sx} dx = 0 \quad \text{for all } s > 0.$$

By the uniqueness theorem $f_\infty(x) = 0$ for $x \geq 0$ as well. That is,

$$f = (f_\infty, f_\ell)^T = 0.$$

Thus, (4.14) implies $f = 0$ and therefore, by Theorem B.2 in Appendix B, the differentiation operator $\dot{D}$ in Case (iii) is a prime symmetric operator as well. □

Remark 4.5. We remark that the symmetric operator $\dot{D}$ in Case (iii) determined by the boundary conditions (4.3) with $k = 0$ is also a prime operator: in this case $\dot{D}$ is unitarily equivalent to the symmetric differential operator in Case (i), which is a prime operator by Lemma 4.4.

Remark 4.6. It is easy to see that the prime symmetric differentiation operator $\dot{D}$ on the metric graph $\mathbb{Y}$ in Cases (i)–(iii) satisfies the semi-Weyl commutation relations in the form (cf. Hypothesis 3.1)

$$U_t^* \dot{D} U_t = \dot{D} + tI \quad \text{on} \quad \text{Dom}(\dot{D}), \quad t \in \mathbb{R}, \tag{4.17}$$

where $U_t = e^{-it\mathcal{Q}}$ is the unitary group generated by the operator $\mathcal{Q}$ of multiplication by independent variable on the graph $\mathbb{Y}$.

To show that the commutation relations (4.17) hold we proceed as follows. Let $\mathcal{A}(x)$ denote a real-valued piecewise continuous function on $\mathbb{Y}$. We remark that the operators $\dot{D}$ and $\dot{D} + \mathcal{A}(x)$ are unitarily equivalent. Indeed, let $\phi(x)$ be any solution to the differential equation

$$\phi'(x) = \mathcal{A}(x) \tag{4.18}$$

on the edges of the graph. Since the graph $\mathbb{Y}$ is a connected tree, the function $\phi(x)$ is determined up to a constant, and we may without loss require that ϕ vanishes at the origin of the graph $\mathbb{Y}$, that is,

$$\phi(0) = 0. \tag{4.19}$$

Denote by V the unitary local gauge transformation

$$(Vf)(x) = e^{i\phi(x)} f(x), \quad f \in L^2(\mathbb{Y}). \tag{4.20}$$

Taking into account the boundary conditions (4.1), (4.2) and (4.3) one concludes that the domain of $\dot{D}$ is V-invariant, that is,

$$V(\text{Dom}(\dot{D})) = \text{Dom}(\dot{D}).$$

Next, a simple computation shows that

$$\dot{D} = V^*(\dot{D} + \mathcal{A}(x))V. \tag{4.21}$$

In the particular case of a constant (magnetic) potential $\mathcal{A}(x) \equiv t$, $t \in \mathbb{R}$, solving (4.18) with the boundary condition (4.19) on the graph $\mathbb{Y}$, one immediately concludes that the unitary operator V from (4.20) is given by

$$V = e^{it\mathcal{Q}},$$

and therefore (4.21) implies the commutation relations (4.17).

Chapter 5

THE MAGNETIC HAMILTONIAN

In this chapter we explicitly describe the set of all self-adjoint (reference) and, more generally, quasi-selfadjoint extensions of the differentiation symmetric operators $\dot{D}$ on the metric graph $\mathbb{Y}$ in Cases (i)–(iii) introduced in Chapter 4.

Theorem 5.1. *Suppose that the metric graph $\mathbb{Y}$ is in one of the Cases (i)-(iii) and let $\dot{D}$ be the symmetric differentiation operator given by (4.1), (4.2) and (4.3), respectively.*

Then the one-parameter family of differentiation operators D_Θ, $|\Theta| = 1$ on the graph $\mathbb{Y}$ in Cases (i)–(iii) with boundary conditions

$$f_\infty(0+) = -\Theta f_\infty(0-), \tag{5.1}$$

$$f_\ell(0) = -\Theta f_\ell(\ell), \tag{5.2}$$

$$\begin{pmatrix} f_\infty(0+) \\ f_\ell(0) \end{pmatrix} = \begin{pmatrix} k & \sqrt{1-k^2}\Theta \\ \sqrt{1-k^2} & -k\Theta \end{pmatrix} \begin{pmatrix} f_\infty(0-) \\ f_\ell(\ell) \end{pmatrix}, \tag{5.3}$$

respectively, coincides with the set of all self-adjoint extensions of the symmetric operator $\dot{D}$.

Moreover, let $g_\pm$ be the deficiency elements of $\dot{D}$ referred to in Lemma 4.3.

Then

$$g = g_+ - \varkappa g_- \in \mathrm{Dom}(D_\Theta), \quad |\varkappa| = 1,$$

if and only if

$$\Theta = F(\varkappa),$$

37

where

$$F(\varkappa) = \begin{cases} \varkappa, & in \ Case \ (i) \\[2mm] -\dfrac{\varkappa - e^{-\ell}}{e^{-\ell}\varkappa - 1}, & in \ Case \ (ii) \\[2mm] -\dfrac{\varkappa - ke^{-\ell}}{ke^{-\ell}\varkappa - 1}, & in \ Case \ (iii). \end{cases} \tag{5.4}$$

Proof. If Υ is in Cases (i)–(ii), the first assertion of the theorem is well known (see, e.g., [3]).

If the graph Υ is in Case (iii) and

$$Y = (-\infty, 0) \sqcup (0, \infty) \sqcup (0, \ell),$$

one can identify the right endpoint of the edge $[0, \ell]$ of the graph Υ with its origin thus making the number of incoming and outgoing bonds equal. Since the incoming $\begin{pmatrix} f_\infty(0-) \\ \ell(\ell) \end{pmatrix}$ and outgoing $\begin{pmatrix} f_\infty(0+) \\ f_\ell(0) \end{pmatrix}$ data are related by the unitary matrix σ with

$$\sigma = \begin{pmatrix} k & \sqrt{1-k^2}\Theta \\ \sqrt{1-k^2} & -k\Theta \end{pmatrix},$$

from [8, Theorem 2.2.1] it follows that the operator D is self-adjoint. Recall that this theorem states that the differentiation operator $D = i\frac{d}{dx}$ on an oriented graph is self-adjoint if and only if for each (finite) vertex v the numbers of incoming and outgoing bonds are equal and the vectors $F^{\text{in}}(v)$ and $F^{\text{out}}(v)$ composed from the values of $f \in \text{Dom}(D)$ attained by f from the incoming and outgoing bonds satisfy the condition

$$F^{\text{out}}(v) = \sigma(v)F^{\text{in}}(v),$$

where σ is a unitary matrix. Next, if $f \in \text{Dom}(\dot{D})$, the boundary conditions (4.3) imply that the boundary conditions (5.3) also hold, and therefore the self-adjoint operator D extends $\dot{D}$.

Conversely, if D is a self-adjoint extension of $\dot{D}$, by [8, Theorem 2.2.1] the boundary conditions

$$\begin{pmatrix} f_\infty(0+) \\ f_\ell(0) \end{pmatrix} = \sigma' \begin{pmatrix} f_\infty(0-) \\ f_\ell(\ell) \end{pmatrix}, \tag{5.5}$$

hold, where σ' is a unitary matrix. The requirement that the self-adjoint operator D extends $\dot{D}$ shows that σ' has to be of the form

$$\sigma' = \begin{pmatrix} k & \alpha \\ \sqrt{1-k^2} & \beta \end{pmatrix}$$

for some α and β. Since σ' is unitary, we have that

$$\sigma' = \begin{pmatrix} k & \sqrt{1-k^2}\Theta \\ \sqrt{1-k^2} & -k\Theta \end{pmatrix}$$

for some Θ, $|\Theta| = 1$, which completes the proof of the first assertion of the theorem.

To prove (5.4) we argue as follows.

We use the following notation $g = g_\infty$ in Case (i), $g = g_\ell$ in Case (ii), and finally, $g = (g_\infty, g_\ell)^T$ in Case (iii) as introduced in Chapter 4.

From the representation (4.6) we get that

$$g_\infty(x) = \sqrt{2}e^x \chi_{(-\infty,0)}(x) - \varkappa\sqrt{2}e^{-x}\chi_{(0,\infty)}(x), \quad x \in \mathbb{R},$$

so that

$$g_\infty(0+) = -\varkappa\sqrt{2} \quad \text{while} \quad g_\infty(0-) = \sqrt{2}.$$

That is, the element g satisfied boundary condition (5.1) with $\Theta = \varkappa$ which proves (5.4) in Case (i).

In Case (ii), we use (4.7) to see that

$$g_\ell(0) = \frac{\sqrt{2}}{\sqrt{e^{2\ell}-1}}(1 - \varkappa e^\ell)$$

and

$$g_\ell(\ell) = \frac{\sqrt{2}}{\sqrt{e^{2\ell}-1}}(e^\ell - \varkappa),$$

which shows that the requirement $g_+ - \varkappa g_- \in \text{Dom}(D_\Theta)$ means that

$$\Theta = -\frac{1 - \varkappa e^\ell}{e^\ell - \varkappa} = -\frac{\varkappa - e^{-\ell}}{\varkappa e^{-\ell} - 1}.$$

In Case (iii), from the representations for the deficiency elements $g_\pm$ (4.8) and (4.9) it follows

$$g_\infty(0+) = a\sqrt{1-k^2}e^\ell\varkappa,$$

$$g_\infty(0-) = a\sqrt{1-k^2},$$

$$g_\ell(\ell) = a(1 - ke^\ell\varkappa),$$

where

$$a = \frac{\sqrt{2}}{\sqrt{e^{2\ell} - k^2}}.$$

Since $g \in \mathrm{Dom}(D_\Theta)$, by (5.3) we have

$$g_\infty(0+) = kg_\infty(0-) + \sqrt{1 - k^2}\Theta g_\ell(\ell),$$

which implies

$$\sqrt{1 - k^2}e^\ell\varkappa = k\sqrt{1 - k^2} + \Theta\sqrt{1 - k^2}(1 - ke^\ell\varkappa).$$

Therefore

$$\Theta = \frac{e^\ell\varkappa - k}{1 - ke^\ell\varkappa} = -\frac{\varkappa - ke^{-\ell}}{ke^{-\ell}\varkappa - 1}. \qquad \Box$$

Remark 5.2. Notice that in Case (i) the self-adjoint operator D_Θ satisfies the semi-Weyl commutation relations

$$U_t^* D_\Theta U_t = D_\Theta + tI \quad \text{on} \quad \mathrm{Dom}(D_\Theta), \quad t \in \mathbb{R}, \quad |\Theta| = 1. \tag{5.6}$$

Here $U_t = e^{-it\mathcal{Q}}$ is the unitary group generated by the self-adjoint operator $\mathcal{Q}$ of multiplication by independent variable on the graph $\mathbb{Y}$.

However, if the graph $\mathbb{Y}$ is in Cases (ii) and (iii), then we only have the commutation relations with respect to a discrete subgroup

$$\mathbb{Z} \ni n \mapsto U_{n\frac{2\pi}{\ell}}$$

of the group U_t. That is,

$$U_{n\frac{2\pi}{\ell}}^* D_\Theta U_{n\frac{2\pi}{\ell}} = D_\Theta + n\frac{2\pi}{\ell}I \quad \text{on} \quad \mathrm{Dom}(D_\Theta), \quad n \in \mathbb{Z}, \quad |\Theta| = 1. \tag{5.7}$$

This phenomenon has the following topological explanation: the set of self-adjoint extensions D_Θ, $|\Theta| = 1$, is in one-to-one correspondence with the unit circle $\mathbb{T}$. The map

$$D_\Theta \to D_{\Theta_t} = U_t^* D_\Theta U_t,$$

determines the flow $\Theta \mapsto \Theta_t$ on $\mathbb{T}$. Using boundary conditions (5.1)-(5.3) it is straightforward to see that

$$\Theta_t = \Theta \begin{cases} 1, & \text{in Case } (i) \\ e^{i\ell t}, & \text{in Cases } (ii) \text{ and } (iii), \end{cases}$$

so that in Cases (ii) and (iii) $\mathrm{Dom}(D_\Theta)$ is not invariant with respect to the whole group U_t, $t \in \mathbb{R}$, but only to its subgroup $U_{n\frac{2\pi}{\ell}}$, $n \in \mathbb{Z}$. In particular, the flow $\Theta \mapsto \Theta_t$ has no fixed point, whenever the graph $\mathbb{Y}$ is in Cases (ii) and (iii) (notice that the Euler characteristic of $\mathbb{T}$ is zero and hence Proposition 3.2 is not applicable). In this regard, it is worth mentioning the fall to the center "catastrophe" in Quantum Mechanics [71, 108, 109]. For a related discussion of the Efimov Effect in three-body systems see [26, 31, 92] where the collapse in a three-body system with point interactions has been discovered, also see [85] and references therein.

More generally, suppose that $\mathcal{A}(x)$ is a real-valued piecewise continuous function on $\mathbb{Y}$. Prescribing the magnetic potential $\mathcal{A}(x)$ to all edges of the graph, consider the *magnetic* differentiation operator $D_\Theta + \mathcal{A}(x)$.

If the graph $\mathbb{Y}$ is in Case (i), the local gauge transformation

$$f(x) \longmapsto e^{i\phi(x)} f(x), \quad f \in L^2(\mathbb{Y}),$$

where $\phi(x)$ is a solution to the differential equation

$$\phi'(x) = \mathcal{A}(x), \quad x \in \mathbb{Y},$$
$$\phi(0) = 0,$$

eliminates the magnetic potential and one shows that the self-adjoint operators D_Θ and $D_\Theta + \mathcal{A}(x)$ are unitarily equivalent, with the unitary equivalence performed by the unitary operator

$$(Vf)(x) = e^{i\phi(x)} f(x), \quad f \in L^2(\mathbb{Y}). \tag{5.8}$$

Clearly $\mathrm{Dom}(D_\Theta)$ is V-invariant, that is,

$$V(\mathrm{Dom}(D_\Theta)) = \mathrm{Dom}(D_\Theta),$$

and therefore

$$D_\Theta = V^*(D_\Theta + \mathcal{A}(x))V. \tag{5.9}$$

If the graph $\mathbb{Y}$ is in Cases (ii) and (iii), the gauge transformation still eliminates the magnetic potential but changes the boundary conditions. That is,

$$V(\mathrm{Dom}(D_\Theta)) = \mathrm{Dom}(D_{\Theta \cdot e^{-i\Phi}}),$$

where

$$\Phi = \int_0^\ell \mathcal{A}(s)ds, \quad \text{the flux of the magnetic field,}$$

and hence

$$D_{\Theta \cdot e^{-i\Phi}} = V^*(D_\Theta + \mathcal{A}(x))V. \tag{5.10}$$

Notice, that in the particular case of a constant potential

$$\mathcal{A}(x) \equiv t,$$

one gets the commutation relations (5.6) and (5.7) as a corollary of (5.9) and (5.10), respectively.

Having in mind the unitary equivalences (5.9) and (5.10) we adopt the following definition.

Definition 5.3. The self-adjoint differentiation operator D_Θ for $|\Theta| = 1$ referred to in Theorem 5.1 will be called the magnetic Hamiltonian.

Notice that in Cases (ii) and (iii), the boundary conditions (5.2) and (5.3) are not local vertex conditions. Bearing in mind applications in quantum mechanics, in Cases (ii) and (iii) one can identify the end points of the interval $[0, \ell]$ to get a one-cycle graph $\overline{\mathbb{Y}}$. As it has been explained in Remark 4.1, in Case (iii) one can also assign two additional vertices to the external edges at $\pm\infty$ of the one-cycle graph $\overline{\mathbb{Y}}$, so that the one-cycle graphs in Case (ii) and (iii) have the Euler characteristic $\chi(\overline{\mathbb{Y}})$ zero with the corresponding first Betti numbers equal to one. In this case, the graph $\overline{\mathbb{Y}}$ can be considered to be the Aharonov-Bohm ring, the configuration space for the quantum system with the magnetic Hamiltonian D_Θ. This system describes a (massless) quantum particle moving on the edges of the graph and the argument of the parameter Θ that determines the magnetic Hamiltonian D_Θ can be interpreted to be the flux of the (zero) magnetic field through the cycle (see (5.10)). For a related information about graphs with Euler characteristic zero in the context of the inverse scattering theory we refer to [67].

Our next goal is to obtain an explicit description of all quasi-selfadjoint extensions of the symmetric differentiation operators $\dot{D}$ introduced in Chapter 4.

Theorem 5.4. *The differentiation operators D_Θ, $\Theta \in \mathbb{C} \cup \{\infty\}$ with $|\Theta| \neq 1$ referred to in Theorem 5.1 with boundary conditions (5.1)–(5.3) is in one to one correspondence with the set of all quasi-selfadjoint extensions of the symmetric operator $\dot{D}$.*

Remark 5.5. If $\Theta = \infty$, the boundary conditions (5.1)–(5.3) in Cases (i)–(iii) should be understood as follows

$$f_\infty(0-) = 0, \tag{5.11}$$

$$f_\ell(\ell) = 0, \tag{5.12}$$

$$\begin{pmatrix} k & \sqrt{1-k^2} \\ 0 & 0 \end{pmatrix} \begin{pmatrix} f_\infty(0+) \\ f_\ell(0) \end{pmatrix} = \begin{pmatrix} f_\infty(0-) \\ f_\ell(\ell) \end{pmatrix}, \tag{5.13}$$

respectively.

Notice that the boundary condition (5.13) can be justified as follows. Rewrite (5.3) as

$$\begin{pmatrix} k & \sqrt{1-k^2}\Theta \\ \sqrt{1-k^2} & -k\Theta \end{pmatrix}^{-1} \begin{pmatrix} f_\infty(0+) \\ f_\ell(0) \end{pmatrix} = \begin{pmatrix} f_\infty(0-) \\ f_\ell(\ell) \end{pmatrix}$$

and observe that

$$\begin{pmatrix} k & \sqrt{1-k^2} \\ 0 & 0 \end{pmatrix} = \lim_{\Theta \to \infty} \begin{pmatrix} k & \sqrt{1-k^2}\Theta \\ \sqrt{1-k^2} & -k\Theta \end{pmatrix}^{-1}$$

$$= \lim_{\Theta \to \infty} \frac{1}{-\Theta} \begin{pmatrix} -k\Theta & -\sqrt{1-k^2}\Theta \\ -\sqrt{1-k^2} & k \end{pmatrix}$$

to get (5.13) as a limiting case.

Notice that the boundary conditions (5.13) can also be rewritten as

$$k f_\infty(0+) + \sqrt{1-k^2} f_\ell(0) = f_\infty(0-),$$

$$f_\ell(\ell) = 0.$$

Proof. If $\mathbb{Y}$ is in Case (i) or (ii), the corresponding result is well known (see, e.g., [3]).

Suppose that the metric graph $\mathbb{Y}$ is in Case (iii). We will describe the required one to one correspondence explicitly.

Denote by $D^\varkappa$ ($\varkappa \in \mathbb{C}$, $|\varkappa| \neq 1$) a quasi-selfadjoint extension of $\dot{D}$ such that

$$\mathrm{Dom}(D^\varkappa) = \mathrm{Dom}(\dot{D}) + \mathrm{span}\{g_+ - \varkappa g_-\}, \tag{5.14}$$

where the deficiency elements $g_\pm$ are given by (4.8) and (4.9).

If

$$f = \begin{pmatrix} f_\infty \\ f_\ell \end{pmatrix} \in \mathrm{Dom}(D^\varkappa), \qquad (5.15)$$

then

$$f_\infty(x) = \alpha\sqrt{1-k^2}e^x\chi_-(x) + \alpha\varkappa\sqrt{1-k^2}e^{\ell-x}\chi_+(x) + h_\infty(x), \quad x \in \mathbb{R},$$

$$f_\ell(x) = \alpha e^x - \alpha\varkappa k e^{\ell-x} + h_\ell(x), \quad x \in [0,\ell),$$

for some $\alpha \in \mathbb{C}$ and some

$$h = \begin{pmatrix} h_\infty \\ h_\ell \end{pmatrix} \in \mathrm{Dom}(\dot{D}).$$

In particular,

$$f_\infty(0-) = \alpha\sqrt{1-k^2} + f_\infty(0-),$$

$$f_\infty(0+) = \alpha\varkappa\sqrt{1-k^2}e^\ell + h_\infty(0+),$$

and

$$f_\ell(0) = \alpha(1 - k\varkappa e^\ell) + h_\ell(0),$$

$$f_\ell(\ell) = \alpha(e^\ell - k\varkappa) + h_\ell(\ell).$$

Since $h \in \mathrm{Dom}(\dot{D})$, the boundary conditions (4.3) hold and therefore

$$f_\infty(0-) = \alpha\sqrt{1-k^2} + h_\infty(0-),$$

$$f_\infty(0+) = \alpha\varkappa\sqrt{1-k^2}e^\ell + kh_\infty(0-),$$

$$f_\ell(0) = \alpha(1 - k\varkappa e^\ell) + \sqrt{1-k^2}h_\infty(0-),$$

$$f_\ell(\ell) = \alpha(e^\ell - k\varkappa).$$

Equivalently,

$$\begin{pmatrix} f_\infty(0+) \\ f_\ell(0) \end{pmatrix} = \begin{pmatrix} \varkappa\sqrt{1-k^2}e^\ell & k \\ 1 - k\varkappa e^\ell & \sqrt{1-k^2} \end{pmatrix} \begin{pmatrix} \alpha \\ h_\infty(0-) \end{pmatrix}$$

and

$$\begin{pmatrix} f_\infty(0-) \\ f_\ell(\ell) \end{pmatrix} = \begin{pmatrix} \sqrt{1-k^2} & 1 \\ e^\ell - k\varkappa & 0 \end{pmatrix} \begin{pmatrix} \alpha \\ h_\infty(0-) \end{pmatrix}.$$

If $e^\ell - k\varkappa \neq 0$, one obtains that

$$\begin{pmatrix} f_\infty(0+) \\ f_\ell(0) \end{pmatrix} = S \begin{pmatrix} f_\infty(0-) \\ f_\ell(\ell) \end{pmatrix}, \tag{5.16}$$

where

$$S = \begin{pmatrix} \varkappa\sqrt{1-k^2}e^\ell & k \\ 1-k\varkappa e^\ell & \sqrt{1-k^2} \end{pmatrix} \begin{pmatrix} \sqrt{1-k^2} & 1 \\ e^\ell - k\varkappa & 0 \end{pmatrix}^{-1}$$

$$= \begin{pmatrix} k & \sqrt{1-k^2}\Theta \\ \sqrt{1-k^2} & -k\Theta \end{pmatrix}.$$

Combining (5.15), (5.16) and (5.3) shows that

$$D^\varkappa = D_\Theta,$$

where

$$\Theta = -\frac{\varkappa e^\ell - k}{k\varkappa - e^\ell}. \tag{5.17}$$

If $e^\ell - k\varkappa = 0$, then necessarily $f_\ell(\ell) = 0$ and

$$kf_\infty(0+) + \sqrt{1-k^2}f_\ell(0) = k\left(\alpha\varkappa\sqrt{1-k^2}e^\ell + kh_\infty(0-)\right)$$

$$+ \sqrt{1-k^2}\left(\alpha(1-k\varkappa e^\ell) + \sqrt{1-k^2}h_\infty(0-)\right)$$

$$= \alpha\sqrt{1-k^2} + h_\infty(0-)$$

$$= f_\infty(0-),$$

which shows that boundary conditions (5.13) holds ($\Theta = \infty$, formally).

It remains to consider the case of the quasi-selfadjoint extension D^∞ defined on

$$\text{Dom}(D^\infty) = \text{Dom}(\dot{D}) + \text{span}\{g_-\}, \tag{5.18}$$

which corresponds to the infinite value of the von Neumann parameter $\varkappa$ ($\varkappa = \infty$).

If (5.18) holds ($\varkappa = \infty$), a similar computation shows that the corresponding quasi-selfadjoint extension corresponds to the boundary condition (5.3) with

$$\Theta = -\frac{e^\ell}{k},$$

which is well defined ($k \neq 0$ by the hypothesis).

Notice that (5.17) gives the link between the boundary condition parameter Θ and the von Neumann extension parameter $\varkappa$ from (5.14) and thus establishes the required correspondence. $\qquad\square$

Remark 5.6. Observe that in Case (i) the operator D_Θ satisfies the semi-Weyl commutation relations

$$U_t^* D_\Theta U_t = D_\Theta + tI \quad \text{on} \quad \text{Dom}(D_\Theta), \quad t \in \mathbb{R}, \tag{5.19}$$

for all $\Theta \in \mathbb{C} \cup \{\infty\}$, where $U_t = e^{-it\mathcal{Q}}$ is the unitary group generated by the self-adjoint operator $\mathcal{Q}$ of multiplication by independent variable on the graph $\mathbb{Y}$.

If the graph $\mathbb{Y}$ is in Cases (ii) and (iii), the commutation relations (5.19) hold only if $\Theta = 0$ or $\Theta = \infty$. Otherwise, we only have the commutation relations with respect to the discrete subgroup $\mathbb{Z} \ni n \mapsto U_{n\frac{2\pi}{\ell}}$, cf. Remark 5.2,

$$U_{n\frac{2\pi}{\ell}}^* D_\Theta U_{n\frac{2\pi}{\ell}} = D_\Theta + n\frac{2\pi}{\ell}I \quad \text{on Dom}(D_\Theta), \quad n \in \mathbb{Z}, \quad |\Theta| = 1. \tag{5.20}$$

It is interesting to notice that the metric graph $\mathbb{Y} = (-\infty, 0) \sqcup (0, \infty) \sqcup (0, \ell)$ in Case (iii) serves as the configuration space for a minimal *self-adjoint dilation* of almost all (with the only one exception) maximal dissipative differentiation operators on the finite interval $(0, \ell)$. Actually, the corresponding self-adjoint dilations coincide with the set of magnetic Hamiltonians D_Θ, $|\Theta| = 1$, in Case (iii).

Theorem 5.7 (cf. [101]). *The self-adjoint operator D_Θ, $|\Theta| = 1$, on the metric graph $\mathbb{Y} = (-\infty, 0) \sqcup (0, \infty) \sqcup (0, \ell)$ in Case (iii) with the boundary conditions (5.3) $(0 < k < 1)$ is a (minimal) self-adjoint dilation of the maximal dissipative differentiation operator $\widehat{d}_\Theta$ on its subgraph $\mathbb{K} = (0, \ell)$ determined by the boundary condition*

$$\text{Dom}(\widehat{d}_\Theta) = \{f_\ell \in W_2^1((0, \ell)) \mid f_\ell(0) = -k\Theta f_\ell(\ell)\}. \tag{5.21}$$

That is,

$$P(D_\Theta - zI)^{-1}|_K = (\widehat{d}_\Theta - zI)^{-1}, \quad z \in \mathbb{C}_-, \tag{5.22}$$

where P is the orthogonal projection from $L^2(\mathbb{Y})$ onto the subspace $K = L^2(\mathbb{K}) = L^2((0, \ell))$. In particular,

$$e^{it\widehat{d}_\Theta} = Pe^{itD_\Theta}|_K, \quad t \geq 0.$$

Proof. Let $g = (g_\infty, g_\ell)^T \in L^2(\mathbb{Y})$ and

$$f = (D_\Theta - zI)^{-1}g, \quad z \in \mathbb{C}_-.$$

Since $f = (f_\infty, f_\ell)^T \in \mathrm{Dom}(D_\Theta)$, the boundary conditions

$$\begin{pmatrix} f_\infty(0+) \\ f_\ell(0) \end{pmatrix} = \begin{pmatrix} k & \sqrt{1-k^2}\Theta \\ \sqrt{1-k^2} & -k\Theta \end{pmatrix} \begin{pmatrix} f_\infty(0-) \\ f_\ell(\ell) \end{pmatrix} \tag{5.23}$$

hold. We have

$$i\frac{d}{dx}f_\infty(x) - zf_\infty(x) = g_\infty(x), \quad x \in (-\infty, 0) \cup (0, \infty) \subset \mathbb{Y},$$

and

$$i\frac{d}{dx}f_\ell(x) - zf_\ell(x) = g_\ell(x), \quad x \in (0, \ell) \subset \mathbb{Y}. \tag{5.24}$$

If $g \in K$, then $g_\infty = 0$ and hence

$$i\frac{d}{dx}f_\infty(x) - zf_\infty(x) = 0, \quad x \in (-\infty, 0) \cup (0, \infty) \subset \mathbb{Y}.$$

Since $z \in \mathbb{C}_-$, the function f_∞ has to vanish on the negative real-axis. In particular, $f_\infty(0-) = 0$. From (5.23) it follows that

$$f_\ell(0) = -k\Theta f_\ell(\ell).$$

Therefore the boundary condition (5.21) holds and hence $f \in \mathrm{Dom}(\widehat{d}_\Theta)$. Combined with (5.24) this means that

$$f_\ell = (\widehat{d}_\Theta - zI)^{-1}g_\ell, \quad z \in \mathbb{C}_-,$$

which proves (5.22) and eventually shows that D_Θ is a self-adjoint dilation of $\widehat{d}_\Theta$. $\square$

Remark 5.8. Theorem 5.7 does not say anything about the dilation of the (exceptional) maximal dissipative differentiation operator $\widehat{d}$ defined on

$$\mathrm{Dom}(\widehat{d}) = \{f_\ell \in W_2^1((0, \ell)) \mid f_\ell(0) = 0\}$$

(it is explicitly assumed that $k \neq 0$ in the boundary condition (5.21)).

In fact, the corresponding self-adjoint dilation coincides with the self-adjoint realization of $i\frac{d}{dx}$ on the metric graph

$$\mathbb{Y} = (-\infty, 0) \sqcup (0, \ell) \sqcup (\ell, \infty),$$

which can be identified with the real axis. Therefore, in the exceptional case the configuration space of the dilation can be identified with the graph $\mathbb{Y}$ in Case (i).

Indeed, to treat the exceptional case, assume that $g \in L^2(\mathbb{R})$ is supported by the finite interval $[0, \ell]$. Then the element

$$f = (D - zI)^{-1}g, \quad z \in \mathbb{C},$$

is supported by the positive semi-axis and its continuous representative satisfies the boundary condition $f(0) = 0$. In particular,

$$i\frac{d}{dx}f(x) - zf(x) = g(x), \quad x \in [0, \ell], \quad f(0) = 0,$$

and therefore the compressed resolvent of D in the lower half-plane coincides with the resolvent of $\hat{d}$ proving that D dilates the dissipative operator $\hat{d}$.

Chapter 6

THE LIVŠIC FUNCTION $s_{(\dot{D}, D_\Theta)}(z)$

The main goal of this and the forthcoming chapter is to describe those unitary invariants of the prime symmetric operator $\dot{D}$ that characterize the operator up to unitary equivalence. Here $\dot{D}$ is the symmetric differentiation operator on the metric graph $\mathbb{Y}$ in one of the Cases (i)–(iii) with boundary conditions (4.1), (4.2) and (4.3), respectively.

To do so, we need to fix a (reference) self-adjoint extension of the operator $\dot{D}$. We choose as such an extension the self-adjoint realization $D = D_\Theta|_{\Theta=1}$ of the differentiation operator referred to in Theorem 5.1. Recall that the domain of the self-adjoint operator $D = D_1$ is characterized by the following boundary conditions

$$f_\infty(0+) = -f_\infty(0-), \tag{6.1}$$

$$f_\ell(0) = -f_\ell(\ell), \tag{6.2}$$

$$\begin{pmatrix} f_\infty(0+) \\ f_\ell(0) \end{pmatrix} = \begin{pmatrix} k & \sqrt{1-k^2} \\ \sqrt{1-k^2} & -k \end{pmatrix} \begin{pmatrix} f_\infty(0-) \\ f_\ell(\ell) \end{pmatrix}, \tag{6.3}$$

whenever the graph $\mathbb{Y}$ is in Cases (i)–(iii), respectively.

We start with the following important observation.

Lemma 6.1. *Let $g_\pm$ be the deficiency elements of the symmetric operator $\dot{D}$ referred to in Lemma 4.3. Then*

$$f = g_+ - g_- \in \mathrm{Dom}(D). \tag{6.4}$$

Proof. In Case (i) we have the representation

$$f(x) = \sqrt{2}(e^x \chi_{(-\infty,0)}(x) - e^{-x}\chi_{(0,\infty)}(x))$$

so that $f(0-) = \sqrt{2} = -f(0+)$ and therefore $f \in \mathrm{Dom}(D)$.

In Case (ii),

$$f(x) = \frac{\sqrt{2}}{\sqrt{e^{2\ell} - 1}}(e^x - e^{\ell-x}), \quad x \in [0, \ell],$$

which implies

$$f(0) = \frac{\sqrt{2}}{\sqrt{e^{2\ell} - 1}}(1 - e^\ell) = -\frac{\sqrt{2}}{\sqrt{e^{2\ell} - 1}}(e^\ell - 1) = -f(\ell)$$

thus showing that $f \in \text{Dom}(D)$ as well.

Finally, in Case (iii), from (4.8) and (4.9) it follows that the element f admits the representation

$$f = \frac{\sqrt{2}}{\sqrt{e^{2\ell} - k^2}}(f_\infty, f_\ell)^T,$$

where

$$f_\infty(x) = \sqrt{1 - k^2}e^x \chi_{(-\infty,0)}(x) + \sqrt{1 - k^2}e^{\ell-x}\chi_{(0,\infty)}(x), \quad x \in \mathbb{R},$$

and

$$f_\ell(x) = e^x - ke^{\ell-x}, \quad x \in [0, \ell].$$

We have

$$f_\infty(0-) = \sqrt{1 - k^2}, \quad f_\infty(0+) = \sqrt{1 - k^2}e^\ell,$$

and

$$f_\ell(0) = 1 - ke^\ell, \quad f_\ell(\ell) = e^\ell - k.$$

Here k, $0 < k < 1$, is the parameter from the boundary conditions (4.3) and (6.3) describing the domains $\text{Dom}(\dot{D})$ and $\text{Dom}(D)$ in Case (iii), respectively.

As a consequence, the incoming $F^{\text{in}} = \begin{pmatrix} f_\infty(0-) \\ f_\ell(\ell) \end{pmatrix}$ and outgoing $F^{\text{in}} = \begin{pmatrix} f_\infty(0+) \\ f_\ell(0) \end{pmatrix}$ boundary data are related as

$$\begin{pmatrix} f_\infty(0+) \\ f_\ell(0) \end{pmatrix} = \begin{pmatrix} \sqrt{1 - k^2}e^\ell \\ 1 - ke^\ell \end{pmatrix} = \begin{pmatrix} k & \sqrt{1 - k^2} \\ \sqrt{1 - k^2} & -k \end{pmatrix} \begin{pmatrix} \sqrt{1 - k^2} \\ e^\ell - k \end{pmatrix}$$

$$= S(1) \begin{pmatrix} f_\infty(0-) \\ f_\ell(\ell) \end{pmatrix},$$

where the bond scattering matrix $S(1)$ is given by (5.3) for $\Theta = 1$, which shows that

$$f \in \text{Dom}(D(1)) = \text{Dom}(D). \qquad \square$$

Based on Lemma 6.1, now we are ready to evaluate the Livšic function associated with the pair $(\dot{D}, D)$, which is one of the unitary invariants that characterizes the pair $(\dot{D}, D)$ up to unitary equivalence.

Lemma 6.2. *The Livšic function associated with the pair* $(\dot{D}, D)$ *admits the representation*

$$s_{(\dot{D}, D)}(z) = \begin{cases} 0, & \text{in Case } (i) \\[2mm] \dfrac{e^{iz\ell} - e^{-\ell}}{e^{-\ell}e^{iz\ell} - 1}, & \text{in Case } (ii) \\[3mm] k\dfrac{e^{iz\ell} - e^{-\ell}}{k^2 e^{-\ell}e^{iz\ell} - 1}, & \text{in Case } (iii). \end{cases} \tag{6.5}$$

Here k, $0 < k < 1$, *is the parameter from the boundary conditions* (4.3) *and* (6.3) *describing the domains* $\mathrm{Dom}(\dot{D})$ *and* $\mathrm{Dom}(D)$ *in Case* (iii).

Proof. Denote by $g_\pm$ the deficiency elements of the symmetric operator $\dot{D}$ referred to in Lemma 4.3. By Lemma 6.1,

$$g_+ - g_- \in \mathrm{Dom}(D) \tag{6.6}$$

in all Cases (i)–(ii). As long as (6.6) is established, in accordance with definition (2.3), the Livšic function associated with the pair $(\dot{D}, D)$ can be evaluated as

$$s_{(\dot{D}, D)}(z) = \frac{z - i}{z + i} \cdot \frac{(g_z, g_-)}{(g_z, g_+)}, \quad z \in \mathbb{C}_+.$$

Here the deficiency elements $g_z \in \mathrm{Ker}((\dot{D})^* - zI)$, $z \in \mathbb{C} \setminus \mathbb{R}$ are given by (4.10), (4.11), (4.12) and (4.13) in Cases (i)–(iii), respectively.

In Case (i), one observes that $g_z \perp g_-$, $z \in \mathbb{C}_+$, and hence

$$s_{(\dot{D}, D)}(z) = 0, \quad z \in \mathbb{C}_+.$$

In Case (ii), one computes

$$s_{(\dot{D}, D)}(z) = \frac{z - i}{z + i} \cdot \frac{(g_z, g_-)}{(g_z, g_+)} = e^\ell \frac{z - i}{z + i} \cdot \frac{\int_0^\ell e^{(-iz-1)x} dx}{\int_0^\ell e^{(-iz+1)x} dx} = e^\ell \frac{e^{(-iz-1)\ell} - 1}{e^{(-iz+1)\ell} - 1}$$

$$= \frac{e^{iz\ell} - e^{-\ell}}{e^{-\ell}e^{iz\ell} - 1}.$$

Finally, in Case (iii), we have

$$s_{(\dot{D},D)}(z) = \frac{z-i}{z+i} \cdot \frac{(g_z, g_-)}{(g_z, g_+)}$$

$$= \frac{z-i}{z+i} \cdot \frac{k \int_0^\ell e^{(-iz-1)x} dx}{(1-k^2) \int_{-\infty}^0 e^{(-iz+1)x} dx + \int_0^\ell e^{(-iz+1)x} dx} e^\ell$$

$$= \frac{k(e^{(-iz-1)\ell} - 1)}{(1-k^2) + e^{(-iz+1)\ell} - 1} e^\ell$$

$$= \frac{k(e^{-iz\ell} - e^\ell)}{e^\ell e^{-iz\ell} - k^2} = k \frac{e^{iz\ell} - e^{-\ell}}{k^2 e^{-\ell} e^{iz\ell} - 1}$$

$$= k \frac{e^{iz\ell} - e^{-\ell}}{k^2 e^{-\ell} e^{iz\ell} - 1}.$$

Combing these results proves (6.5).   □

Remark 6.3. The representation (6.5) in Cases (i) and (ii) is known (see, e.g., [3]).

The following corollary provides a complete characterization of prime symmetric operators with deficiency indices $(1,1)$ satisfying the commutation relations (1.3) (see Problem (I) a) in the Introduction).

Corollary 6.4. *Let $\dot{A}$ be a symmetric operator referred to in Hypothesis 3.1. Suppose that $\dot{A}$ is a prime operator. Then $\dot{A}$ is unitarily equivalent to one of the differentiation operators $\dot{D} = i\frac{d}{dx}$ on a metric graph $\mathbb{Y}$ in one of the Cases (i)–(iii) introduced in Chapter 4 (see eqs. (4.1), (4.2) and (4.3)).*

Proof. By Corollary 3.7, $\dot{A}$ admits a self-adjoint extensions such that the Livšic function associated with the pair $(\dot{A}, A)$ coincides with the one referred to in Lemma 6.2. Since $\dot{A}$ is a prime operator, by the Uniqueness Theorem C.1 in Appendix C, the operator $\dot{A}$ is unitarily equivalent to the symmetric differentiation operator on the metric graph $\mathbb{Y}$ in one of the cases Cases (i)–(iii).   □

More generally, the Livšic function associated with the pair $(\dot{D}, D_\Theta)$, $|\Theta| = 1$, where D_Θ is the self-adjoint realization of the differentiation operator referred to in Theorem 5.1, differs from the function $s_{(\dot{D},D)}(z)$ evaluated above in Lemma 6.2 by a constant unimodular factor. For the sake of completeness, we present the following result.

Theorem 6.5. *Suppose that* $|\Theta| = 1$. *The Livšic function* $s_{(\dot{D},D_\Theta)}(z)$ *associated with the pair* $(\dot{D}, D_\Theta)$ *admits the representation*

$$s_{(\dot{D},D_\Theta)}(z) = e^{-2i\alpha} \begin{cases} 0, & in \ Case \ (i) \\ \dfrac{e^{iz\ell} - e^{-\ell}}{e^{-\ell}e^{iz\ell} - 1}, & in \ Case \ (ii) \\ k\dfrac{e^{iz\ell} - e^{-\ell}}{k^2 e^{-\ell}e^{iz\ell} - 1}, & in \ Case \ (iii). \end{cases} \quad (6.7)$$

Here k, $0 < k < 1$, *is the parameter from the boundary conditions* (4.3) *and* (5.3) *describing the domains* $\mathrm{Dom}(\dot{D})$ *and* $\mathrm{Dom}(D_\Theta)$ *in Case* (iii), *respectively.*

Here α *and the boundary condition parameter* Θ *are related as follows*

$$e^{2i\alpha} = \begin{cases} \Theta, & in \ Case \ (i) \\ \dfrac{\Theta + e^{-\ell}}{e^{-\ell}\Theta + 1}, & in \ Case \ (ii) \\ \dfrac{\Theta + e^{-(\ell+\ell')}}{e^{-(\ell+\ell')}\Theta + 1}, & in \ Case \ (iii) \end{cases} \quad , \quad with \ \ell' = \ln\frac{1}{k}. \quad (6.8)$$

In particular,

$$s_{(\dot{D},D_\Theta)}(z) = e^{-2i\alpha} s_{(\dot{D},D)}(z), \quad (6.9)$$

where

$$D = D_\Theta|_{\Theta=1}.$$

Proof. From Theorem 5.1 it follows that

$$F = g_+ - e^{2i\alpha}g_- \in \mathrm{Dom}(D_\Theta), \quad (6.10)$$

where $g_\pm$ are the deficiency elements referred to in Lemma 4.3 and $e^{2i\alpha}$ is given by (6.8). Now (6.7) follows from (6.5) by Lemma E.1 in Appendix E. $\qquad\square$

We conclude this chapter with several remarks of analytic character.

Remark 6.6. (i) One observes that the Livšic function

$$s^{II}(z; \ell) = \frac{e^{iz\ell} - e^{-\ell}}{e^{-\ell}e^{iz\ell} - 1}, \quad z \in \mathbb{C}_+,$$

given by (6.7) in Case (ii) admits the representations

$$s^{II}(z; \ell) = \frac{e^{i\ell z} - e^{-\ell}}{e^{-\ell}e^{i\ell z} - 1} = \frac{\sin(z - i)\frac{\ell}{2}}{\sin(z + i)\frac{\ell}{2}}$$

$$= \frac{z - i}{z + i} \cdot \frac{\prod\limits_{n \in \mathbb{Z}} \left(1 - \left(\frac{z-i}{2\pi n}\ell\right)^2\right)}{\prod\limits_{n \in \mathbb{Z}} \left(1 - \left(\frac{z+i}{2\pi n}\ell\right)^2\right)}.$$

Therefore, $s^{II}(z; \ell)$ is a pure Blaschke product with zeroes z_n located on the lattice

$$z_n = i + \frac{2\pi}{\ell}n, \quad n \in \mathbb{Z}.$$

(ii) A direct computation shows that the Livšic function

$$s^{III}(z; k, \ell) = k\frac{e^{iz\ell} - e^{-\ell}}{k^2 e^{-\ell}e^{iz\ell} - 1}, \quad z \in \mathbb{C}_+, \quad 0 < k < 1,$$

(cf. (6.7) in Case (iii)) can be obtained by an analytic continuation of $s^{II}(z; \ell)$ with an appropriate identification of the parameters. That is,

$$s^{III}(z; e^{-\ell'}, \ell) = s^{II}\left(\frac{\ell z + i\ell'}{\ell + \ell'}; \ell + \ell'\right), \quad z \in \mathbb{C}_+. \quad (6.11)$$

(iii) In the inner-outer factorization of the Livšic function in Case (iii)

$$s^{III}(z; k, \ell) = k\frac{e^{iz\ell} - e^{-\ell}}{k^2 e^{-\ell}e^{iz\ell} - 1} = s_{\text{in}}^{III}(z) \cdot s_{\text{out}}^{III}(z)$$

the inner factor $s_{\text{in}}^{III}(z)$ coincides with the Livšic function in Case (ii), i.e.,

$$s_{\text{in}}^{III} = s^{II}(z; \ell) = \frac{e^{iz\ell} - e^{-\ell}}{e^{-\ell}e^{iz\ell} - 1}.$$

Indeed,

$$s^{III}(z; e^{-\ell'}, \ell) = \frac{\sin(z - i)\dfrac{\ell}{2}}{\sin\left((z + i)\dfrac{\ell}{2} + i\ell'\right)}$$

$$= s^{II}(z; \ell) \cdot \frac{\sin(z + i)\dfrac{\ell}{2}}{\sin\left((z + i)\dfrac{\ell}{2} + i\ell'\right)}.$$

(In particular, the functions $s^{III}(z; e^{-\ell'}, \ell)$ and $s^{II}(z; \ell)$ have the same set of zeros).

To complete the proof of the claim it remains to show that the function

$$t(z) = \frac{\sin(z + i)\dfrac{\ell}{2}}{\sin\left((z + i)\dfrac{\ell}{2} + i\ell'\right)} \tag{6.12}$$

is an outer function.

First, one observes that $t(z)$ is a contractive function in the upper half-plane. Next, let

$$t(z) = t_{\text{in}}(z) t_{\text{out}}(z)$$

be its inner-outer factorization. Since $t(z)$ does not vanish in the upper half-plane, the inner factor of $t(z)$ is necessarily a singular inner function. Since $t(z)$ admits an analytic continuation into a strip in the lower half-plane, the singular measure in the exponential representation of $t_{\text{in}}(z)$ does not charge bounded sets and therefore

$$t_{\text{in}}(z) = e^{iLz}$$

for some $L \geq 0$, "mass" at infinity. In particular,

$$\lim_{y \to \infty} t_{\text{in}}(iy) = 0$$

unless $L = 0$. However, from (6.12) it follows that

$$\lim_{y \to \infty} t(iy) = e^{-\ell'},$$

which implies that $L = 0$. Therefore $t_{\text{in}}(z) = 1$ and hence $t(z)$ is an outer function.

(iv) In fact, for the outer factor $s_{out}^{III}(z) = t(z)$ one gets the representation

$$s_{out}^{III}(z) = \sqrt{\frac{\sinh \ell}{\sinh(\ell + 2\ell')}}$$

$$\times \exp\left(\frac{i}{2\pi} \int_{\mathbb{R}} \left(\frac{1}{\lambda + z} + \frac{\lambda}{1 + \lambda^2}\right) \rho(\lambda) d\lambda\right), \qquad (6.13)$$

where the density is given by

$$\rho(\lambda) = \log \frac{P_{e^{-\ell - 2\ell'}}(\lambda\ell)}{P_{e^{-\ell}}(\lambda\ell)}.$$

Here

$$P_r(\theta) = \frac{1 - r^2}{1 + r^2 - 2r \cos \theta}$$

is the Poisson kernel.

Indeed, since $t(z)$ is an outer function in the upper half-plane, we have the representation [60]

$$t(z) = \exp\left(\frac{i}{\pi} \int_{\mathbb{R}} \left(\frac{1}{\lambda + z} + \frac{\lambda}{1 + \lambda^2}\right) \log |t(\lambda)| d\lambda\right). \qquad (6.14)$$

Using (6.12) one computes that

$$\log |t(\lambda)| = \log \left| e^{-\ell'} \frac{e^{iz\ell} - e^{-\ell}}{e^{-2\ell'} e^{-\ell} e^{iz\ell} - 1}\right|$$

$$= -\ell' + \frac{1}{2} \log \frac{(\cos \ell\lambda - e^{-\ell})^2 + \sin^2 \ell\lambda}{(e^{-\ell - 2\ell'} \cos \ell\lambda - 1)^2 + e^{-2\ell - 4\ell'} \sin^2 \ell\lambda}$$

$$= -\ell' + \frac{1}{2} \log \frac{1 - 2\cos \ell\lambda e^{-\ell} + e^{-2\ell}}{1 - 2e^{-\ell - 2\ell'} \cos \ell\lambda + e^{-2\ell - 4\ell'}}$$

$$= \frac{1}{2} \log \left(e^{-2\ell'} \cdot \frac{1 - e^{-2\ell}}{1 - e^{-2\ell - 4\ell'}} \cdot \frac{P_{e^{-\ell - 2\ell'}}(\lambda\ell)}{P_{e^{-\ell}}(\lambda\ell)}\right)$$

$$= \frac{1}{2} \log \frac{\sinh \ell}{\sinh(\ell + 2\ell')} + \rho(\lambda),$$

and since

$$\frac{i}{\pi} \int_{\mathbb{R}} \left(\frac{1}{\lambda + z} + \frac{\lambda}{1 + \lambda^2}\right) d\lambda = 1, \quad z \in \mathbb{C}_+,$$

representation (6.14) simplifies to (6.13).

Chapter 7

THE WEYL-TITCHMARSH FUNCTION $M_{(\dot{D},D_\Theta)}(z)$

Along with the Livšic function, the Weyl-Titchmarsh function associated with a pair consisting of a prime symmetric operator and its self-adjoint extension characterizes the pair up to mutual unitary equivalence. So that our next goal is to evaluate the Weyl-Titchmarsh function associated with the symmetric differentiation $\dot{D}$ on the metric graph $\mathbb{Y}$ and its self-adjoint reference extension.

Suppose that $\mathbb{Y}$ is the metric graph $\mathbb{Y}$ in one of the Cases (i)–(iii). As it follows from Lemma 6.2, the Weyl-Titchmarsh function $M_{(\dot{D},D)}$ associated with the pair $(\dot{D}, D)$ has the form

$$
M_{(\dot{D},D)}(z) = \begin{cases} i, & \text{in Case (i)} \\[2mm] \coth \dfrac{\ell}{2} \tan \dfrac{\ell}{2} z, & \text{in Case (ii)} \\[2mm] \coth \dfrac{\ell + \ell'}{2} \tan \left(\dfrac{\ell}{2} z + i \dfrac{\ell'}{2} \right), & \text{in Case (iii)} \end{cases} , \qquad (7.1)
$$

where

$$
\ell' = \ln \frac{1}{k} \quad (0 < k < 1)
$$

and k, $0 < k < 1$, is the parameter from the boundary conditions (4.3) and (6.3) describing the domains $\mathrm{Dom}(\dot{D})$ and $\mathrm{Dom}(D)$ in Case (iii).

Indeed, in Case (iii) one observes that

$$M_{(\dot{D},D)}(z) = \frac{1}{i} \frac{s_{(\dot{D},D)}(z) + 1}{s_{(\dot{D},D)}(z) - 1} = \frac{1}{i} \frac{\frac{ke^{iz\ell} - ke^{-\ell}}{k^2 e^{-\ell} e^{iz\ell} - 1} + 1}{\frac{ke^{iz\ell} - ke^{-\ell}}{k^2 e^{-\ell} e^{iz\ell} - 1} - 1}$$

$$= \frac{1}{i} \cdot \frac{1 + ke^{-\ell}}{1 - ke^{-\ell}} \cdot \frac{ke^{i\ell z} - 1}{ke^{i\ell z} + 1} = \coth \frac{\ell + \ell'}{2} \cdot \tan\left(\frac{\ell}{2}z + i\frac{\ell'}{2}\right).$$

Case (ii) then follows by setting $\ell' = 0$, equivalently $k = 1$, and the corresponding representation in Case (i) is obvious (formally take the limit as $\ell \to \infty$ in Case (ii)).

More generally, we have the following result.

Theorem 7.1. *Let* D_Θ, $|\Theta| = 1$, *be the one-parameter family of self-adjoint reference operators referred to in Theorem 5.1. Then the Weyl-Titchmarsh function* $M_{(\dot{D},D_\Theta)}$ *associated with the pair* $(\dot{D}, D_\Theta)$ *admits the representation*

$$M_{(\dot{D},D_\Theta)}(z) = \begin{cases} i, & \text{in Case } (i) \\[2mm] A(\Phi)\tan\left(\frac{\ell}{2}z - \frac{\Phi}{2}\right) + \frac{\sin\Phi}{\sinh\ell}, & \text{in Case } (ii) \\[2mm] A(\Phi)\tan\left(\frac{\ell}{2}z + \frac{\ell'}{2}i - \frac{\Phi}{2}\right) + \frac{\sin\Phi}{\sinh(\ell + \ell')}, & \text{in Case } (iii) \end{cases},$$

$$(7.2)$$

where

$$\Phi = \arg\Theta, \quad \ell' = \log\frac{1}{k} \quad (0 < k < 1)$$

and k, $0 < k < 1$, *is the parameter from the boundary conditions* (4.3) *and* (5.3) *describing the domains* $\operatorname{Dom}(\dot{D})$ *and* $\operatorname{Dom}(D_\Theta)$ *in Case* (iii).

Here in Case (iii) *the amplitude* $A(\Phi)$ *is given by the convex combination*

$$A(\Phi) = \cos^2\frac{\Phi}{2} \cdot \coth\frac{\ell + \ell'}{2} + \sin^2\frac{\Phi}{2} \cdot \tanh\frac{\ell + \ell'}{2}, \qquad (7.3)$$

and in Case (ii) $A(\Phi)$ *is given by the same expression with* $\ell' = 0$.

Proof. In Case (i) there is nothing to prove, since $s_{(\dot{D},D_\Theta)}(z) = 0$ and hence $M_{(\dot{D},D_\Theta)}(z) = i$ for all $z \in \mathbb{C}_+$.

In Case (ii), by Theorem 6.5 (see eq. (6.9)), we have

$$s_{(\dot{D},D_\Theta)}(z) = e^{-2i\alpha} s_{(\dot{D},D)}(z), \qquad (7.4)$$

where

$$D = D_\Theta|_{\Theta=1}$$

and α and Θ are related as in (6.8).

From (7.4) and Lemma E.1 in Appendix E it follows

$$M_{(\dot{D}, D_\Theta)}(z) = \frac{M_{(\dot{D}, D)}(z) - \tan\alpha}{1 + \tan\alpha \cdot M_{(\dot{D}, D)}(z)}. \tag{7.5}$$

By (7.1),

$$M_{(\dot{D}, D)}(z) = \coth\frac{\ell}{2}\tan\frac{\ell}{2}z = m\tan\zeta =: \mathcal{M}_{(\dot{D}, D)}(\zeta), \tag{7.6}$$

where

$$m = \coth\frac{\ell}{2} \quad \text{and} \quad \zeta = \frac{\ell}{2}z, \tag{7.7}$$

and therefore

$$M_{(\dot{D}, D_\Theta)}(z) = \frac{m\tan\zeta - \tan\alpha}{1 + m\tan\alpha\tan\zeta}.$$

Using the identity

$$\frac{m\tan\zeta - \tan\alpha}{1 + m\tan\alpha\tan\zeta} = \frac{1 + \tan^2\alpha}{1 + m^2\tan^2\alpha}m\frac{\tan\zeta - m\tan\alpha}{1 + m\tan\alpha\tan\zeta}$$
$$+ \frac{m^2\tan\alpha - \tan\alpha}{1 + m^2\tan^2\alpha},$$

we obtain the following representation

$$M_{(\dot{D}, D_\Theta)}(z) = \frac{1 + \tan^2\alpha}{1 + m^2\tan^2\alpha}\mathcal{M}_{(\dot{D}, D)}(\zeta - t)$$
$$+ \frac{m^2\tan\alpha - \tan\alpha}{1 + m^2\tan^2\alpha}, \tag{7.8}$$

where

$$\tan t = m\tan\alpha.$$

Therefore, (7.8) can be rewritten as

$$M_{(\dot{D}, D_\Theta)}(z) = \frac{1 + \frac{1}{m^2}\tan^2 t}{1 + \tan^2 t}\mathcal{M}_{(\dot{D}, D)}(\zeta - t) + \frac{m - \frac{1}{m}}{1 + \tan^2 t}\tan t.$$

In view of (7.6) and (7.7), we have

$$M_{(\dot{D},D_\Theta)}(z) = \left(\coth\frac{\ell}{2}\cos^2 t + \tanh\frac{\ell}{2}\sin^2 t\right)$$

$$\times \tan\left(\frac{\ell}{2}z - t\right) + \frac{1}{\sinh\ell}\sin 2t. \qquad (7.9)$$

From (6.8) it follows that

$$\tan\alpha = \frac{1}{i}\frac{\frac{\Theta - e^{-\ell}}{e^{-\ell}\Theta - 1} - 1}{\frac{\Theta - e^{-\ell}}{e^{-\ell}\Theta - 1} + 1} = \tanh\frac{\ell}{2}\cdot\tan\left(\frac{1}{2}\arg\Theta\right),$$

so that

$$\tan t = m\tan\alpha = \coth\frac{\ell}{2}\cdot\tan\alpha = \tan\left(\frac{1}{2}\arg\Theta\right).$$

In particular,

$$t = \frac{1}{2}\arg\Theta = \frac{1}{2}\Phi, \qquad (7.10)$$

which along with (7.9) shows that

$$M_{(\dot{D},D_\Theta)}(z) = A(\Phi)\tan\left(\frac{\ell}{2}z - \frac{\Phi}{2}\right) + \frac{\sin\Phi}{\sinh\ell},$$

proving (7.2) with $A(\Phi)$ given by (7.3) in Case (ii).

To prove (7.2) in Case (iii), in a similar way one gets (cf. (7.5))

$$M_{(\dot{D},D)}(z) = \coth\frac{\ell + \ell'}{2}\tan\left(\frac{\ell}{2}z + i\frac{\ell'}{2}\right)$$

and establishes that (7.8) holds where now

$$\zeta = \frac{\ell}{2}z + i\frac{\ell'}{2}, \quad m = \coth\frac{\ell + \ell'}{2}$$

and

$$\tan t = m\tan\alpha.$$

Observing that

$$\tan\alpha = \frac{1}{i}\frac{\frac{\Theta + e^{-\ell - \ell'}}{e^{-\ell - \ell'}\Theta + 1} - 1}{\frac{\Theta + e^{-\ell - \ell'}}{e^{-\ell - \ell'}\Theta + 1} + 1} = \tanh\frac{\ell + \ell'}{2}\tan\left(\frac{1}{2}\arg\Theta\right),$$

one justifies (7.10) in Case (iii) as well. Literally repeating the reasoning above one obtains the representation (7.2) in Case (iii). $\qquad\square$

Our last result in this chapter shows that the spectral measure of the reference operator D_Θ, $|\Theta| = 1$, is rather sensitive to the magnitude of the "flux" $\Phi = \arg \Theta$.

Corollary 7.2. *The Weyl-Titchmarsh function $M_\Theta(z) = M_{(\dot{D}, D_\Theta)}(z)$ associated with the pair $(\dot{D}, D_\Theta)$ admits the representation*

$$M_\Theta(z) = \int_{\mathbb{R}} \left(\frac{1}{\lambda - z} - \frac{\lambda}{1 + \lambda^2} \right) d\mu_\Theta(\lambda), \quad |\Theta| = 1,$$

where $\mu_\Theta(d\lambda)$, the spectral measure, is

(i) *the absolutely continuous measure with a constant density*

$$\mu_\Theta(d\lambda) = \frac{1}{\pi} d\lambda \quad \text{in Case (i)}; \tag{7.11}$$

(ii) *the discrete pure point measure*

$$\mu_\Theta(d\lambda) = \frac{2}{\ell} A(\arg \Theta) \sum_{k \in \mathbb{Z}} \delta_{\frac{(2k+1)\pi + \Phi}{\ell}}(d\lambda) \quad \text{in Case (ii)}, \tag{7.12}$$

with $\delta_x(d\lambda)$ the Dirac mass at x and

$$A(\Phi) = \cos^2 \frac{\Phi}{2} \cdot \coth \frac{\ell}{2} + \sin^2 \frac{\Phi}{2} \cdot \tanh \frac{\ell}{2};$$

(iii) *the absolutely continuous measure with a periodic density*

$$\mu_\Theta(d\lambda) = \frac{1}{\pi} A(\arg \Theta) P_{e^{-\ell'}} (\ell\lambda - \pi - \arg \Theta) d\lambda. \tag{7.13}$$

Here

$$P_r(\varphi) = \frac{1 - r^2}{1 + r^2 - 2r \cos \varphi}$$

is the Poisson kernel,

$$\ell' = \log \frac{1}{k},$$

with k, $0 < k < 1$, the parameter from the boundary conditions (4.3), (5.3), and

$$A(\Phi) = \cos^2 \frac{\Phi}{2} \cdot \coth \frac{\ell + \ell'}{2} + \sin^2 \frac{\Phi}{2} \cdot \tanh \frac{\ell + \ell'}{2}.$$

Proof. Indeed, since

$$i = \frac{1}{\pi} \int_{\mathbb{R}} \left(\frac{1}{\lambda - z} - \frac{\lambda}{1 + \lambda^2} \right) d\lambda, \quad z \in \mathbb{C}_+,$$

(7.11) follows from the equality $M_\Theta = i$ in Case (i).

To check (7.12), we use the representation

$$\tan(z) = -\sum_{k=0}^{\infty} \left(\frac{1}{z - (k + \frac{1}{2})\pi} + \frac{1}{z + (k + \frac{1}{2})\pi} \right) \tag{7.14}$$

$$= \int_{\mathbb{R}} \left(\frac{1}{\lambda - z} - \frac{\lambda}{1 + \lambda^2} \right) d\nu(\lambda),$$

where $\nu(d\lambda)$ is a discrete point measure

$$\nu(d\lambda) = \sum_{k \in \mathbb{Z}} \delta_{(k + \frac{1}{2})\pi}(d\lambda), \tag{7.15}$$

which proves (7.12) in the particular case of $\ell = 2$, and $\Theta = 1$, and then the general case follows by making a simple change of variables.

Using the explicit representation (7.2) for the Weyl-Titchmarsh function $M_\Theta(z)$ in Case (iii) one observes that $M_\Theta(z)$ is bounded in the upper half-plane. Therefore, the representing measure $\mu_\Theta(d\lambda)$ in Case (iii) is absolutely continuous with the density given by

$$\mu_\Theta(d\lambda) = \frac{1}{\pi} \operatorname{Im} M_\Theta(\lambda + i0) d\lambda = \frac{1}{\pi} A(\arg \Theta)$$

$$\cdot \operatorname{Im} \tan \left(\frac{\ell\lambda + \ell'i - \arg \Theta}{2} \right) d\lambda$$

$$= \frac{1}{\pi} A(\arg \Theta) P_{e^{-\ell'}} (\ell\lambda - \pi - \arg \Theta) \, d\lambda, \tag{7.16}$$

which proves (7.13).

Here we have used the representation for the imaginary part of the tangent function of a complex argument via the the Poisson kernel,

$$\operatorname{Im} \tan (\lambda + i\tau) = \operatorname{Im} \frac{1}{i} \cdot \frac{e^{i(\lambda + i\tau)} - e^{-i(\lambda + i\tau)}}{e^{i(\lambda + i\tau)} + e^{-i(\lambda + i\tau)}}$$

$$= -\operatorname{Re} \frac{e^{i\lambda}e^{-\tau} - e^{-i\lambda}e^{\tau}}{e^{i\lambda}e^{-\tau} + e^{-i\lambda}e^{\tau}} \cdot \frac{e^{-i\lambda}e^{-\tau} + e^{i\lambda}e^{\tau}}{e^{-i\lambda}e^{-\tau} + e^{i\lambda}e^{\tau}}$$

$$= \frac{e^{2\tau} - e^{-2\tau}}{e^{2\tau} + 2\cos 2\lambda + e^{-2\tau}} = P_{e^{-2\tau}}(2\lambda - \pi),$$

$$\lambda \in \mathbb{R}, \quad \tau > 0. \qquad \square$$

Remark 7.3. Observing that

$$\frac{1}{2\pi} P_r(\lambda) d\lambda \to \sum_n \delta_n(d\lambda) \quad \text{as} \quad r \uparrow 1,$$

we get the following spectral hierarchy of the representing measures given by (7.13), (7.12) and (7.11):

$$\frac{1}{\pi} A(\arg \Theta) \cdot P_{e^{-\ell'}} (\ell \lambda - \pi - \arg \Theta) \, d\lambda$$

$$\downarrow \quad (\ell' \to 0)$$

$$\frac{2}{\ell} A(\arg \Theta) \cdot \sum_{k \in \mathbb{Z}} \delta_{\frac{(2\pi k + 1)\pi + \Phi}{\ell}} (d\lambda)$$

$$\downarrow \quad (\ell \to \infty)$$

$$\frac{1}{\pi} d\lambda,$$

with the limits as $\ell' \to 0$ and $\ell \to \infty$ taken in the sense of the weak convergence of the measures. Here we used the inequality (see (7.3))

$$\tanh \frac{\ell + \ell'}{2} \leq A(\Phi) \leq \coth \frac{\ell + \ell'}{2}$$

on the last step that ensures that the amplitude $A(\Phi)$ approaches 1 as $\ell \to \infty$.

Chapter 8

THE MODEL DISSIPATIVE OPERATORS

Given a metric graph $\mathbb{Y}$ in one of the Cases (i)–(iii) and a real parameter k, $0 \leq k < 1$, we construct a family of *model* maximal dissipative differentiation operators on $\mathbb{Y}$. In what follows we will refer to the parameter k as *the quantum gate coefficient on the graph* $\mathbb{Y}$.

If the metric graph $\mathbb{Y}$ is in Case (i),

$$\mathbb{Y} = (-\infty, 0) \sqcup (0, \infty),$$

denote by

$$\widehat{D} = \widehat{D}_I(k) = i\frac{d}{dx}, \quad 0 \leq k < 1,$$

the maximal dissipative differentiation operator with the boundary condition at the origin

$$f_\infty(0+) = k f_\infty(0-), \quad 0 \leq k < 1. \tag{8.1}$$

Notice that the case $k = 0$ is exceptional in the sense that the point spectrum of the dissipative operator $\widehat{D}_I(0)$ fills in the entire (open) upper half-plane.

The corresponding strongly continuous semi-group generated by $\widehat{D}_I(k)$ describes the motion from left to right of a (quantum) particle which is emitted outside (see Fig. 8.1) with probability $1 - k^2$ through the quantum gate at the origin and keeps moving along the axes with probability k^2.

If the metric graph $\mathbb{Y}$ is in Case (ii) and

$$\mathbb{Y} = (0, \ell)$$

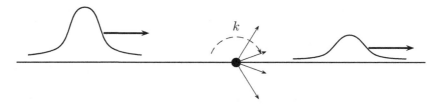

Fig. 8.1. Dynamics on the Metric Graph $\mathbb{Y}$ in Case (i) with the Quantum Gate Coefficient k

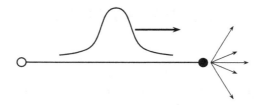

Fig. 8.2. Dynamics on the Metric Graph $\mathbb{Y}$ in Case (ii) with the Quantum Gate Coefficient $k = 0$

for some $\ell > 0$, denote by

$$\widehat{D} = \widehat{D}_{II}(k, \ell) = i\frac{d}{dx}, \tag{8.2}$$

the maximal dissipative differentiation operator determined by the boundary condition

$$f_\ell(0) = k f_\ell(\ell), \quad 0 \leq k < 1. \tag{8.3}$$

Notice that the case $k = 0$ is also exceptional. That is, the dissipative differentiation operator $\widehat{D}_{II}(0, \ell)$ corresponding to the boundary condition

$$f_\ell(0) = 0 \tag{8.4}$$

has no spectrum.

The (nilpotent) semi-group generated by $\widehat{D}_{II}(0, \ell)$ describes the motion of a particle which is emitted with probability one at the right end-point of the finite interval $[0, \ell]$ (see Fig. 8.2).

Finally, if the graph $\mathbb{Y}$ is in Case (iii),

$$\mathbb{Y} = (-\infty, 0) \sqcup (0, \infty) \sqcup (0, \ell),$$

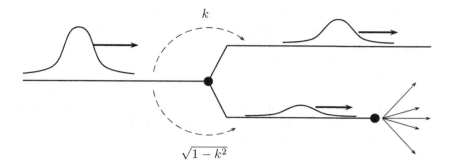

Fig. 8.3. Dynamics on the Metric Graph $\mathbb{Y}$ in Case (iii) with the Quantum Gate Coefficient k

denote by $\widehat{D} = \widehat{D}_{III}(k,\ell) = i\frac{d}{dx}$ the maximal dissipative differentiation operator on $\mathbb{Y}$ with the boundary conditions

$$\begin{pmatrix} f_\infty(0+) \\ f_\ell(0) \end{pmatrix} = \begin{pmatrix} k & 0 \\ \sqrt{1-k^2} & 0 \end{pmatrix} \begin{pmatrix} f_\infty(0-) \\ f_\ell(\ell) \end{pmatrix}, \quad 0 < k < 1. \tag{8.5}$$

The dynamics associated with the strongly continuous semi-group generated by $\widehat{D}_{III}(k,\ell)$ describes the motion a wave-packet moving from left to right (see Fig. 8.3).

If the packet is initially supported by the negative semi-axis, after the interaction with the scatterer located at the center of the graph, the particle continues its rightward motion along the real axis with its initial shape amplified by the factor k while a copy of the wave-packet amplified by $\sqrt{1-k^2}$ turns right onto an appendix of length ℓ attached to the obstacle. When the wave-packet approaches the right end of the interval $[0,\ell]$ the wave is terminated.

From the boundary conditions (8.5) it follows that the quantum Kirchhoff rule (at the junction)

$$|f_\infty(0-)|^2 = |f_\infty(0+)|^2 + |f_\ell(0)|^2 \tag{8.6}$$

holds.

Taking into account wave-particle duality, one can also say that the corresponding particle with probability k^2 keeps moving along the real axis and with probability $1 - k^2$ enters the appendix. Then, assuming that the initial profile of the wave-packet was supported by the interval $[-L,0)$, the particle is emitted with probability one after time $t = \ell + L$ has elapsed.

Notice that a wave-packet initially supported to the right of the obstacle moves to the right without changing its shape regardless whether the wave is supported by the semi-axis $(-\infty, 0]$ or by the finite interval $[0, \ell]$. To complete the description of the dynamics in the general case, one applies the superposition principle.

Remark 8.1. Notice that the boundary conditions (8.1), (8.4) and (8.5) (but not (8.3) with $k \neq 0$) are the local vertex conditions, which means that different vertices do not interact. In particular, the domains of the corresponding dissipative operators $\widehat{D}$ are invariant with respect to the group of local gauge transformations. As a corollary, the dissipative operators $\widehat{D} = \widehat{D}_{I,II,III}$ satisfy the commutation relations

$$U_t^* \widehat{D} U_t = \widehat{D} + tI \quad \text{on} \quad \text{Dom}(\widehat{D}), \quad t \in \mathbb{R}, \tag{8.7}$$

where $U_t = e^{-it\mathcal{Q}}$ is the unitary group generated by the operator $\mathcal{Q}$ of multiplication by independent variable on the graph $\mathbb{Y}$.

This can be justified immediately but it also follows from a more general considerations below (cf. Remark 4.6).

Let $\mathcal{A}(x)$ denote a real-valued piecewise continuous function on $\mathbb{Y}$. We remark that the operators $\widehat{D}$ and $\widehat{D} + \mathcal{A}(x)$ are unitarily equivalent. Indeed, let $\phi(x)$ be any solution to the differential equation

$$\phi'(x) = \mathcal{A}(x),$$

on the edges of the graph and continuous at the origin $\{0\} \in \mathbb{Y}$. Denote by V the unitary local gauge transformation

$$(Vf)(x) = e^{i\phi(x)} f(x), \quad f \in L^2(\mathbb{Y}). \tag{8.8}$$

Then, taking into account the boundary conditions (8.1), (8.4) and (8.5) one concludes that the domain of $\widehat{D}$ is V-invariant

$$V(\text{Dom}(\widehat{D})) = \text{Dom}(\widehat{D}),$$

and moreover,

$$\widehat{D} = V^*(\widehat{D} + \mathcal{A}(x))V.$$

In particular, (8.7) holds.

Remark 8.2. Notice that the model dissipative differentiation operators $\widehat{D}$ extend the symmetric differentiation operator $\dot{D}$ on the graph $\mathbb{Y}$ with the boundary conditions (4.1), (4.2) and (4.3), respectively. Moreover, the symmetric operator $\dot{D}$ is uniquely determined by $\widehat{D}$ and

$$\dot{D} = \widehat{D}|_{\mathcal{D}} \quad \text{where} \quad \mathcal{D} = \text{Dom}(\widehat{D}) \cap \text{Dom}((\widehat{D})^*).$$

If $\widehat{D}$ is in Case (i), then the corresponding symmetric operator $\dot{D}$ admits a quasi-selfadjoint extension the point spectrum of which fills in the entire upper half-plane. Notice that this property characterizes the operator up to unitary equivalence. That is, any prime closed symmetric operator with deficiency indices $(1, 1)$ that admits a quasi-selfadjoint extension with point spectrum filling in the whole upper half-plane is unitarily equivalent to the operator $\dot{D}$ in Case (i) [3, Ch. IX, Sec. 114]. Apparently, any point from $\mathbb{C}_+$ is an eigenvalue for the extension $\widehat{D}_I(0)$.

Moreover, the symmetric operator $\dot{D}$ has a relatively poor family of unitarily inequivalent (dissipative) quasi-selfadjoint extensions. The reason is that any two (dissipative) extensions with the same absolute value of the von Neumann parameter k, $(0 \le k \le 1)$, are unitarily equivalent to the operator $\widehat{D}_I(k)$. Recall that the absolute value of the von Neumann parameter of the dissipative operators in question is well defined (see Remark 2.5). This property also characterizes $\dot{D}$ up to unitary equivalence: any prime closed symmetric operator with deficiency indices $(1, 1)$ that admits two distinct unitarily equivalent quasi-selfadjoint extensions is unitarily equivalent to the operator $\dot{D}$ [5, Theorem 2].

The dissipative operator $\widehat{D} = \widehat{D}_{II}(0, \ell)$ in Case (ii) given by (8.2) and (8.3) is the only dissipative extension of $\dot{D}_{II}(\ell)$ whose resolvent set coincides with the whole complex plane $\mathbb{C}$. Moreover, any dissipative quasi-selfadjoint extension of a prime closed symmetric operator with deficiency indices $(1, 1)$ without spectrum is unitarily equivalent to the symmetric differentiation operator $\dot{D}_{II}(\ell)$ on a finite interval of length ℓ [73, Theorem 14].

We remark that in contrast to Case (i), in Cases (ii) and (iii) the maximal dissipative operator $\widehat{D}$ with the boundary conditions (8.1) and (8.4), respectively, is the only one maximal dissipative extension of the symmetric differentiation operator $\dot{D}$ with a gauge invariant domain. Indeed, the boundary conditions (5.2) and (5.3) are gauge invariant if either $\Theta = 0$ or $\Theta = \infty$. If $\Theta = \infty$, the corresponding quasi-selfadjoint extension is not dissipative, which proves the claim.

The following structure theorem shows that the differentiation operator $\widehat{D}$,

$$\widehat{D} = \widehat{D}_{III}(k, \ell),$$

in Case (iii) can be obtained as the result of an operator coupling (spectral synthesis) of the more "elementary" dissipative differentiation operators $\widehat{D} = \widehat{D}_I(k)$ in Case (i) and $\widehat{D} = \widehat{D}_{II}(0,\ell)$ in Case (ii). For the concept of an operator coupling we refer to Appendix G.

Theorem 8.3. *The differentiation operator* $\widehat{D}_{III}(k,\ell)$ *on the metric graph*

$$\mathbb{Y}_{III} = (-\infty, 0] \sqcup [0, \infty) \sqcup [0, \ell]$$

with the quantum gate coefficient k *is an operator coupling of the differentiation operator* $\widehat{D}_I(k)$ *on the edge*

$$\mathbb{Y}_I = (-\infty, 0] \sqcup [0, \infty)$$

with the same quantum gate coefficient k *and the operator* $\widehat{D}_{II}(0,\ell)$ *on the remaining edge*

$$\mathbb{Y}_{II} = [0, \ell]$$

with the quantum gate coefficient 0, *respectively. That is,*

$$\mathbb{Y}_{III} = \mathbb{Y}_I \sqcup \mathbb{Y}_{II}$$

and

$$\widehat{D}_{III}(k,\ell) = \widehat{D}_I(k) \uplus \widehat{D}_{II}(0,\ell). \tag{8.9}$$

Proof.

To see that, set $\mathcal{H}_1 = L^2((-\infty, \infty))$ and $\mathcal{H}_2 = L^2((0, \ell))$. One observes that

$$\widehat{D}_{III}(k,\ell)\big|_{\mathrm{Dom}(\widehat{D}_{III}(k,\ell)) \cap \mathcal{H}_1} = \widehat{D}_I(k)$$

and hence the requirement (i) in the definition G.4 (see Appendix G) of a coupling of two dissipative operators is met.

Next, $\mathrm{Dom}(\widehat{D}_I(0,k)) \oplus \mathrm{Dom}((\widehat{D}_{II}(\ell))^*)$ consists of the three-component functions $f = (f_-, f_+, f_\ell)^T$, with

$$f_- \oplus f_+ \oplus f_\ell \in W_2^1((-\infty, 0)) \oplus W_2^1((0, \infty)) \oplus W_2^1((0, \ell))$$

such that

$$f_+(0+) = kf_-(0-) \quad \text{and} \quad f_\ell(\ell-) = 0 \tag{8.10}$$

and hence

$$\mathrm{Dom}(\dot{D}) \subset \mathrm{Dom}(\widehat{D}_I(k)) \oplus \mathrm{Dom}((\widehat{D}_{II}(0,\ell))^*).$$

In particular,

$$\dot{D} \subset \widehat{D}_I(k) \oplus (\widehat{D}_{II}(0,\ell))^*$$

and hence the requirement (ii) in the definition of an operator coupling is met as well. Therefore, (8.9) holds. □

For the further references it is convenient to adopt the following Hypothesis.

Hypothesis 8.4. Assume that the metric graph Υ is in one of the Cases (i)–(iii) and $\widehat{D}$ is the model dissipative operator with the quantum gate coefficient k on Υ given by the boundary conditions (8.1), (8.3) and (8.5), respectively. Let $\dot{D}$ be the restriction of $\widehat{D}$ on $\mathcal{D} = \mathrm{Dom}(\widehat{D}) \cap \mathrm{Dom}((\widehat{D})^*)$. Assume that D_Θ is the self-adjoint reference extension of $\dot{D}$ referred to in Theorem 5.1 (see (5.1), (5.2), and (5.3)). If the graph Υ is in Case (ii) assume that $k = 0$ and if Υ is in Case (iii) we require that $k \neq 0$.

Definition 8.5. Under Hypothesis 8.4 suppose that $k = 0$ whenever the graph Υ is in Case (ii) and that $k \neq 0$ whenever Υ is in Case (iii). Under this assumption we call the triple of differentiation operators $(\dot{D}, \widehat{D}, D_\Theta)$ *the model triple on Υ with the quantum gate coefficient k.*

More explicitly, each of the differentiation operators from the triple $(\dot{D}, \widehat{D}, D_\Theta)$ in the Hilbert $L^2(\Upsilon)$ is given by the differential expression

$$\tau = i\frac{d}{dx}$$

(on the edges of the graph Υ) initially defined on the Sobolev space $W_2^1(\Upsilon)$,

$$W_2^1(\Upsilon) = \begin{cases} W_2^1((-\infty,0)) \oplus W_2^1((0,\infty)) \\ W_2^1((0,\ell)) \\ (W_2^1((-\infty,0)) \oplus W_2^1((0,\infty)) \oplus W_2^1((0,\ell)). \end{cases}$$

In Case (i), the metric graph has the form $\Upsilon = (-\infty,0) \sqcup (0,\infty)$, and

$$\mathrm{Dom}(\dot{D}) = \{f_\infty \in W_2^1(\Upsilon) \,|\, f_\infty(0+) = f_\infty(0-) = 0\},$$

$$\mathrm{Dom}(\widehat{D}) = \{f_\infty \in W_2^1(\Upsilon) \,|\, f_\infty(0+) = kf_\infty(0-)\},$$

$$\mathrm{Dom}(D_\Theta) = \{f_\infty \in W_2^1(\Upsilon) \,|\, f_\infty(0+) = -\Theta f_\infty(0-)\},$$

$$0 \leq k < 1.$$

In Case (ii), $\mathbb{Y} = (0, \ell)$,

$$\mathrm{Dom}(\dot{D}) = \{f_\ell \in W_2^1(\mathbb{Y}) \,|\, f_\ell(0) = f_\ell(\ell) = 0\},$$

$$\mathrm{Dom}(\widehat{D}) = \{f_\ell \in W_2^1(\mathbb{Y}) \,|\, f_\ell(0) = 0\},$$

$$\mathrm{Dom}(D_\Theta) = \{f_\ell \in W_2^1(\mathbb{Y}) \,|\, f_\ell(0) = -\Theta f_\ell(\ell)\}.$$

In Case (iii), $\mathbb{Y} = (-\infty, 0) \sqcup (0, \infty) \sqcup (0, \ell)$, and

$$\mathrm{Dom}(\dot{D}) = \left\{ f_\infty \oplus f_\ell \in W_2^1(\mathbb{Y}) \,\Big|\, \begin{cases} f_\infty(0+) = k f_\infty(0-) \\ f_\ell(0+) = \sqrt{1-k^2} f_\infty(0-) \\ f_\ell(\ell) = 0 \end{cases} \right\},$$

$$\mathrm{Dom}(\widehat{D}) = \left\{ f_\infty \oplus f_\ell \in W_2^1(\mathbb{Y}) \,\Big|\, \begin{cases} f_\infty(0+) = k f_\infty(0-) \\ f_\ell(0+) = \sqrt{1-k^2} f_\infty(0-) \end{cases} \right\},$$

$$\mathrm{Dom}(D_\Theta) = \left\{ f_\infty \oplus f_\ell \in W_2^1(\mathbb{Y}) \,\Big|\, \begin{cases} f_\infty(0+) = k f_\infty(0-) + \sqrt{1-k^2}\,\Theta f_\ell(\ell) \\ f_\ell(0+) = \sqrt{1-k^2} f_\infty(0-) - k\Theta f_\ell(\ell) \end{cases} \right\},$$

$$0 < k < 1.$$

Chapter 9

THE CHARACTERISTIC FUNCTION $S_{(\dot{D},\widehat{D},D_\Theta)}(z)$

Now we are ready to evaluate the characteristic function $S_{(\dot{D},\widehat{D},D_\Theta)}(z)$ of the triple $(\dot{D},\widehat{D},D_\Theta)$ of differentiation operators on a metric graph $\mathbb{Y}$ in Cases (i)–(iii).

First, we evaluate the characteristic function of the triple $(\dot{D},\widehat{D},D)$ for a particular choice of the reference self-adjoint operator D referred to in Theorem 5.1, in Cases (i), (ii) and (iii) with $\Theta = 1$, respectively. Recall that the operator D is determined by the boundary conditions

$$f_\infty(0+) = -f_\infty(0-),$$

$$f_\ell(0) = -f_\ell(\ell),$$

$$\begin{cases} f_\infty(0+) &= k f_\infty(0-) + \sqrt{1-k^2} f_\ell(\ell) \\ f_\ell(0+) &= \sqrt{1-k^2} f_\infty(0-) - k f_\ell(\ell) \end{cases}$$

in Cases (i)–(iii), respectively.

Lemma 9.1. *Let $(\dot{D},\widehat{D},D)$ be the model triple of the differentiation operators on the metric graph $\mathbb{Y}$ in one of the Cases (i)–(iii) as above. Then the characteristic function of the triple $(\dot{D},\widehat{D},D)$ admits the representation*

$$S_{(\dot{D},\widehat{D},D)}(z) = \begin{cases} k, & \text{in Case (i)} \\ e^{i\ell z}, & \text{in Case (ii)} \\ k e^{i\ell z}, & \text{in Case (iii)}, \end{cases} \quad z \in \mathbb{C}_+, \tag{9.1}$$

where $0 \le k < 1$ (in Case (i)) and $0 < k < 1$ (in Case (iii)).

Proof. To check (9.1) we proceed as follows.

Let $g_\pm$ be the deficiency elements given by (4.6) in Case (i), (4.7) in Case (ii) and (4.8), (4.9) in Case (iii).

We claim that

$$f = g_+ - S(i)g_- \in \text{Dom}(\widehat{D}), \tag{9.2}$$

where we have used the shorthand notation

$$S(z) = S_{(\dot{D},\widehat{D},D)}(z).$$

It suffices to check that f satisfies the boundary conditions (8.1), (8.4) and (8.5) in Cases (i)–(iii), respectively, and therefore

$$f = g_+ - S(i)g_- \in \text{Dom}(\widehat{D}).$$

Indeed, in Case (i),

$$S(i) = k, \tag{9.3}$$

and hence

$$f_\infty(x) = \sqrt{2}(e^x \chi_{(-\infty,0)}(x) + k e^{-x} \chi_{(0,\infty)}(x)).$$

Clearly,

$$f_\infty(0-) = k f_\infty(0+),$$

and therefore $f_\infty \in \text{Dom}(\widehat{D})$.

In Case (ii),

$$S(i) = e^{-\ell}, \tag{9.4}$$

and therefore the element

$$f_\ell(x) = \frac{\sqrt{2}}{\sqrt{e^{2\ell} - 1}}(e^x - e^{-\ell}e^{\ell-x}), \quad x \in [0, \ell],$$

satisfies the Dirichlet boundary condition $f_\ell(0) = 0$ which proves that $f_\ell \in \text{Dom}(\widehat{D})$.

Finally, in Case (iii) we have that

$$S(i) = k e^{-\ell}. \tag{9.5}$$

Therefore, the element $f = g_+ - S(i)g_- = g_+ - ke^{-\ell}g_-$ admits the representation (see (4.8) and (4.9))

$$f = \frac{\sqrt{2}}{\sqrt{e^{2\ell} - k^2}}(f_\infty, f_\ell)^T,$$

where

$$f_\infty(x) = \sqrt{1 - k^2}\, e^x \chi_{(-\infty,0)}(x) + ke^{-\ell}\sqrt{1 - k^2}\, e^{\ell-x}\chi_{(0,\infty)}(x), \quad x \in \mathbb{R},$$

and

$$f_\ell(x) = e^x - (ke^{-\ell})ke^{\ell-x}, \quad x \in [0, \ell].$$

We have

$$f_\infty(0-) = \sqrt{1 - k^2},$$

$$f_\infty(0+) = k\sqrt{1 - k^2},$$

$$f_\ell(0) = 1 - k^2,$$

which shows that the boundary conditions (8.5) hold. Therefore $f \in \text{Dom}(\hat{D})$.

Since

$$g_+ - g_- \in \text{Dom}(D)$$

and

$$g_+ - S(i)g_- \in \text{Dom}(\hat{D}),$$

one computes

$$S_{(\dot{D},\hat{D},D)}(z) = \frac{s_{(\dot{D},D)}(z) - S(i)}{S(i)s_{(\dot{D},D)}(z) - 1} = \frac{s_{(\dot{D},D)}(z) - ke^{-\ell}}{ke^{-\ell}s_{(\dot{D},D)}(z) - 1}.$$

It remains to remark that since the Livšic function $s_{(\dot{D},D)}(z)$ is given by (6.5), one gets (9.1) by a direct computation. $\qquad\square$

Remark 9.2. Notice that Lemma 9.1, in particular, states that the characteristic function of the triple $(\dot{D}, \hat{D}, D)$ in Case (iii) is the product of the characteristic functions in Cases (i) and (ii), respectively. That is,

$$S_{(\dot{D},\hat{D},D)}(z) = k \cdot e^{i\ell z}. \tag{9.6}$$

In view of Theorem 8.3, the rule (9.6) can also be obtained as a corollary of the Multiplication Theorem G.5 in Appendix G.

Also, comparing (9.3), (9.4) and (9.5), one observes that the von Neumann parameter $ke^{-\ell}$ of the coupling $\widehat{D}_{III}(k,\ell)$ associated with the bases (4.8) and (4.9) is the product of the von Neumann parameters k and $e^{-\ell}$ of the dissipative operators $\widehat{D}_I(k)$ and $\widehat{D}_{II}(\ell)$ with respect to the bases (4.6) and (4.7), respectively (see Remark 2.1 for the terminology). This observation illustrates the multiplicativity property for the absolute values of the von Neumann extension parameters under coupling (see [87, Theorem 5.4] or Theorem G.5 in Appendix G). Recall that the concept of absolute value of the von Neumann parameter is well defined by Remark 2.5.

The more general result below can be understood as the solution of the following inverse problem: find a triple with a prescribed characteristic function referred to in Theorem 3.5 (cf. [55, Theorem 20]).

Theorem 9.3. *The characteristic function* $S_{(\dot{D},\widehat{D},D_\Theta)}(z)$ *of the model triple* $(\dot{D}, \widehat{D}, D_\Theta)$, $|\Theta| = 1$, *admits the representation*

$$S_{(\dot{D},\widehat{D},D_\Theta)}(z) = e^{-2i\alpha} \begin{cases} k, & in \ Case \ (i) \\ e^{i\ell z}, & in \ Case \ (ii) \\ ke^{i\ell z}, & in \ Case \ (iii) \end{cases} , \quad z \in \mathbb{C}_+, \qquad (9.7)$$

where $0 \le k < 1$ *(in Case (i)) and* $0 < k < 1$ *(in Case (iii)).*

Here α *and the boundary condition parameter* Θ *are related as follows*

$$e^{2i\alpha} = \begin{cases} \Theta, & in \ Case \ (i) \\ \dfrac{\Theta + e^{-\ell}}{e^{-\ell}\Theta + 1}, & in \ Case \ (ii) \\ \dfrac{\Theta + e^{-(\ell+\ell')}}{e^{-(\ell+\ell')}\Theta + 1}, & in \ Case \ (iii), \quad with \ \ell' = \ln \frac{1}{k}. \end{cases} \qquad (9.8)$$

In particular,

$$S_{(\dot{D},\widehat{D},D_\Theta)}(z) = e^{-2i\alpha} S_{(\dot{D},\widehat{D},D)}(z). \qquad (9.9)$$

Proof. In view of Lemma E.1 in Appendix E, the assertion of the theorem is a direct consequence of Theorem 6.5.   □

Remark 9.4. Observe that in Case (ii) the characteristic function of the triple $(\dot{D}, \widehat{D}, D_\Theta)$ is a singular inner function with "mass at infinity."

Corollary 9.5. *The model triples* $(\dot{D}, \widehat{D}, D_\Theta)$, $|\Theta| = 1$, *and* $(\dot{D}, \widehat{D}, D_{\Theta'})$, $|\Theta'| = 1$, $\Theta' \ne \Theta$, *are not mutually unitarily equivalent unless the graph*

Υ *is in Case (i) and the point spectrum of the dissipative differentiation operator $\widehat{D}$ fills in the whole upper half-plane $\mathbb{C}_+$. In the latter case the triples in question are mutually unitarily equivalent to one another for any Θ and Θ' ($|\Theta| = |\Theta'| = 1$).*

Proof. Combining (9.7), (9.8) and (9.9) shows that the characteristic functions $S_{(\dot{D},\widehat{D},D_\Theta)}(z)$ and $S_{(\dot{D},\widehat{D},D_{\Theta'})}(z)$ are different for $\Theta \neq \Theta'$ unless

$$S_{(\dot{D},\widehat{D},D_\Theta)}(z) = 0 \quad \text{for all } z \in \mathbb{C}_+$$

for some (and therefore for all) $|\Theta| = 1$. In the latter case $\widehat{D}$ has no regular points in the upper half-plane and the triples $(\dot{D}, \widehat{D}, D_\Theta)$, $|\Theta| = 1$, and $(\dot{D}, \widehat{D}, D_{\Theta'})$, $|\Theta'| = 1$, are mutually unitarily equivalent by the uniqueness Theorem C.1 in Appendix C. □

Remark 9.6. Let $\mathcal{A}(x)$ be a real-valued piecewise continuous function on the metric graph Υ in Cases (i)–(iii). Combining Remarks 4.6, 5.2 and 8.1 imply that if the graph Υ is in Case (i), then the triple $(\dot{D} + \mathcal{A}(x), \widehat{D} + \mathcal{A}(x), D_\Theta + \mathcal{A}(x))$ is mutually unitarily equivalent to the triple $(\dot{D}, \widehat{D}, D_\Theta)$. If the graph is in Cases (ii) or (iii), then $(\dot{D} + \mathcal{A}(x), \widehat{D} + \mathcal{A}(x), D_\Theta + \mathcal{A}(x))$ is mutually unitarily equivalent to the triple $(\dot{D}, \widehat{D}, D_{\Theta e^{-i\Phi}})$, where

$$\Phi = \int_0^\ell \mathcal{A}(x)dx.$$

Moreover, the corresponding unitary equivalence is given by a gauge transformation.

The knowledge of the characteristic function of the triple $(\dot{D}, \widehat{D}, D_\Theta)$, which is its complete unitary invariant, enables us to obtain the converse to structure Theorem 3.5.

Theorem 9.7. *Let $(\dot{A}, \widehat{A}, A)$ be a triple of operators in a Hilbert space $\mathcal{H}$ where $\dot{A}$ is a prime symmetric operator with deficiency indices $(1,1)$, $\widehat{A}$ is its dissipative quasi-selfadjoint extension and A is a reference self-adjoint extension of $\dot{A}$.*

Suppose that the characteristic function $S(z) = S_{(\dot{A},\widehat{A},A)}(z)$ of the triple $(\dot{A}, \widehat{A}, A)$ admits the representation

$$S(z) = ke^{i\ell z}, \quad z \in \mathbb{C}_+, \tag{9.10}$$

for some $|k| \leq 1$ and $\ell \geq 0$. We also assume that if $\ell = 0$, then necessarily $|k| < 1$ and if $|k| = 1$, then $\ell > 0$.

Then there exists a unitary group U_t such that the domains $\mathrm{Dom}(\dot{A})$ and $\mathrm{Dom}(\widehat{A})$ are U_t-invariant and

$$U_t^* \dot{A} U_t = \dot{A} + tI \quad \text{on } \mathrm{Dom}(\dot{A}), \tag{9.11}$$

$$U_t^* \widehat{A} U_t = \widehat{A} + tI \quad \text{on } \mathrm{Dom}(\widehat{A}). \tag{9.12}$$

Proof. By the uniqueness Theorem C.1 in Appendix C, the triple $(\dot{A}, \widehat{A}, A)$ is mutually unitarily equivalent to the triple $(\dot{D}, \widehat{D}, D_\Theta)$ referred to in Theorem 9.3 for some choice of the extension parameter Θ. By (4.17) and (8.7),

$$e^{it\mathcal{Q}} \dot{D} e^{-it\mathcal{Q}} = \dot{D} + tI \quad \text{on } \mathrm{Dom}(\dot{D}), \quad t \in \mathbb{R},$$

and

$$e^{it\mathcal{Q}} \widehat{D} e^{-it\mathcal{Q}} = \widehat{D} + tI \quad \text{on } \mathrm{Dom}(\widehat{D}), \quad t \in \mathbb{R},$$

where $\mathcal{Q}$ the self-adjoint operator of multiplication by independent variable on the graph $\mathbb{Y}$. Pulling back the group $e^{-it\mathcal{Q}}$ in $L^2(\mathbb{Y})$ to the original Hilbert space $\mathcal{H}$ proves the assertion. $\qquad\square$

It is worth mentioning that the choice of the orientation of the graph $\mathbb{Y}$ was *ad hoc* from the very beginning. To complete the exposition, along with the graph $\mathbb{Y}$, consider the metric graph $\mathbb{Y}^*$ obtained from $\mathbb{Y}$ by reversing the orientation. The corresponding differentiation operator $-(\widehat{D})^*$ on $\mathbb{Y}^*$ extends the (symmetric) differentiation operator

$$-\dot{D} = -i\frac{d}{dx},$$

and its domain is determined by the following boundary conditions
in Case (i):

$$f_\infty(0-) = kf_\infty(0+); \tag{9.13}$$

in Case (ii):

$$f_\ell(\ell) = 0; \tag{9.14}$$

in Case (iii):

$$\begin{cases} f_\infty(0-) &= kf_\infty(0+) + \sqrt{1-k^2}f_\ell(0) \\ f_\ell(\ell) &= 0 \end{cases}. \tag{9.15}$$

Notice that the graph $\mathbb{Y}^*$ in Case (iii) (as opposed to the graph $\mathbb{Y}$) has only two incoming and only one outgoing bonds which is reflected in a slightly different way of posing boundary conditions (cf. (8.5) and (9.15)). Meanwhile, both $-(\dot{D})^*$ and $-(\widehat{D})^*$ solve the commutation relations (8.7). On the algebraic level, it can be seen by observing that the relations

$$U_t^* \widehat{A} U_t = \widehat{A} + tI \quad \text{on } \mathrm{Dom}(\widehat{A}),$$

and

$$V_t^* (-\widehat{A})^* V_t = (-\widehat{A})^* + tI \quad \text{on } \mathrm{Dom}(\widehat{A}),$$

with $V_t = U_{-t}$, imply one another.

In fact, reversing the orientation of the graph $\mathbb{Y}$ does not lead to the new solutions as far as the classification up to unitary equivalence is concerned.

Lemma 9.8. *The triples $(\dot{D}, \widehat{D}, D_\Theta)$ and $(-\dot{D}, -(\widehat{D})^*, -D_{\overline{\Theta}})$ are mutually unitarily equivalent.*

Proof. From Theorem 9.3 it follows that

$$S_{(\dot{D},\widehat{D},D_\Theta)}(z) = e^{-2i\alpha} k e^{i\ell z},$$

where α is given by (6.8). Applying Lemma F.2 in Appendix F, we have that

$$S_{(-\dot{D},-(\widehat{D})^*,-D_{\overline{\Theta}})}(z) = \overline{S_{(\dot{D},\widehat{D},D_{\overline{\Theta}})}(-\overline{z})} = e^{-2i\alpha} k e^{i\ell z} = S_{(\dot{D},\widehat{D},D_\Theta)}(z),$$

which ensures a mutual unitary equivalence of the triples in question by the uniqueness Theorem C.1 in Appendix C (the symmetric operator $\dot{D}$ is prime by Lemma 4.4, so is the operator $-\dot{D}$). $\qquad\square$

Chapter 10

THE TRANSMISSION COEFFICIENT AND THE CHARACTERISTIC FUNCTION

Recall that if the metric graph $\mathbb{Y}$ is in Cases (ii) or (iii), the differentiation operators D_Θ, $\Theta \in \mathbb{C} \cup \{\infty\}$, referred to in Theorem 5.4 satisfy the commutation relations

$$U_t^* D_\Theta U_t = D_\Theta + tI, \quad t \in \frac{2\pi}{\ell}\mathbb{Z},$$

with respect to a discrete subgroup of one-parameter strongly continuous group of unitary operators U_t. On fact, one can choose

$$U_t = e^{-it\mathcal{Q}}, \quad t \in \mathbb{R},$$

where $\mathcal{Q}$ is the multiplication operator by independent variable on the graph $\mathbb{Y}$. However, in the exceptional cases $\Theta = \infty$, the semi-Weyl relations

$$U_t^* D_\Theta U_t = D_\Theta + tI, \quad t \in \mathbb{R}, \quad (\Theta = 0 \quad \text{or} \quad \Theta = \infty),$$

hold $(D_\infty = -D_0^*)$.

Suppose that the metric graph $\mathbb{Y}$ is in Case (ii), that is,

$$\mathbb{Y} = (0, \ell).$$

Our fist goal is to evaluate the characteristic function of a dissipative triple on the graph $\mathbb{Y}$.

To be more specific, let $(\dot{d}, \widehat{d}_\Theta, d)$ be the triple, where $\dot{d}$ is the symmetric differentiation on

$$\mathrm{Dom}(\dot{d}) = \{f_\ell \in W_2^1((0, \ell)) \mid f_\ell(0) = f_\ell(\ell) = 0\},$$
$$\widehat{d}_\Theta = \widehat{D}_{II}(-k\Theta, \ell) \quad (0 < k < 1)$$

is the maximal dissipative differentiation operator on

$$\mathrm{Dom}(\widehat{d}_\Theta) = \{f_\ell \in W_2^1((0, \ell)) \mid f_\ell(0) = -k\Theta f_\ell(\ell)\},$$

and d is the self-adjoint differentiation operator defined on

$$\mathrm{Dom}(d) = \{f_\ell \in W_2^1((0, \ell)) \mid f_\ell(0) = -f_\ell(\ell)\}.$$

Lemma 10.1. *The characteristic function* $S_{(\dot{d}, \widehat{d}_\Theta, d)}(z)$ *of the triple* $(\dot{d}, \widehat{d}_\Theta, d)$ *has the form*

$$S_{(\dot{d}, \widehat{d}_\Theta, d)}(z) = \frac{\Theta + e^{-\ell}k}{ke^{-\ell}\Theta + 1} \cdot B(z), \quad z \in \mathbb{C}_+, \tag{10.1}$$

where

$$B(z) = \frac{e^{i\ell z} + k\Theta}{\Theta + e^{i\ell z}k} = \frac{\cos \frac{\ell z - \arg \Theta - i\ell'}{2}}{\cos \frac{\ell z - \arg \Theta + i\ell'}{2}}, \quad z \in \mathbb{C}_+,$$

is a pure Blaschke product with simple zeros z_n *given by*

$$z_n = i\frac{\ell'}{\ell} + (2\pi n + 1)\frac{\pi}{\ell} + \frac{\arg \Theta}{\ell}, \quad \text{with} \quad \ell' = \log \frac{1}{k}, \tag{10.2}$$
$$0 < k < 1.$$

In particular, the characteristic function $S_{(\dot{d}, \widehat{d}_\Theta, d)}(z)$ *of the triple* $(\dot{d}, \widehat{d}_\Theta, d)$ *is a periodic function*

$$S_{(\dot{d}, \widehat{d}_\Theta, d)}\left(z + \frac{2\pi}{\ell}\right) = S_{(\dot{d}, \widehat{d}_\Theta, d)}(z), \quad z \in \mathbb{C}_+,$$

with the minimal period $T = \frac{2\pi}{\ell}$.

Proof. Let $g_\pm$ be the deficiency elements of the symmetric operator $\dot{d}$ given by (4.7).

Since $\widehat{d}_\Theta$ is a quasi-selfadjoint extension of $\dot{d}$, its domain can be represented as

$$\mathrm{Dom}(\widehat{d}_\Theta) = \mathrm{Dom}(\dot{d})\dotplus\mathrm{span}\{g_+ - \varkappa g_-\},$$

for some $|\varkappa| < 1$. In particular the function $g_+ - \varkappa g_-$ satisfies the boundary condition

$$(g_+ - \varkappa g_-)(0) = -k\Theta(g_+ - \varkappa g_-)(\ell),$$

which allows to relate the von Neumann extension parameter $\varkappa$ and the coefficient k as

$$1 - \varkappa e^\ell = -k\Theta(e^\ell - \varkappa).$$

Therefore,

$$\varkappa = \frac{k\Theta + e^{-\ell}}{k\Theta e^{-\ell} + 1}.$$

Since $g_+ - g_- \in \mathrm{Dom}(d)$, one computes (see (2.9)) that

$$S(z) = S_{(d,\widehat{d}_\Theta,d)}(z) = \frac{s(z) - \varkappa}{\overline{\varkappa}s(z) - 1}, \tag{10.3}$$

where $s(z) = s_{(\dot{d},d)}(z)$ is the Livšic function associated with the pair $(\dot{d}, d)$ given by Lemma 6.2 as

$$s(z) = s_{(\dot{d},d)}(z) = \frac{e^{i\ell z} - e^{-\ell}}{e^{-\ell}e^{i\ell z} - 1}.$$

We claim that the inner functions $B(z)$ and $S(z)$ have the same set of (simple) roots.

Indeed, if $S(z_0) = 0$, then

$$s(z_0) = \frac{e^{i\ell z_0} - e^{-\ell}}{e^{-\ell}e^{i\ell z_0} - 1} = \varkappa. \tag{10.4}$$

Therefore,

$$e^{i\ell z_0} = \frac{e^{-\ell} - \varkappa}{1 - \varkappa e^{-\ell}} = \frac{1 - e^\ell \varkappa}{e^\ell - \varkappa} = -k\Theta,$$

which implies that $B(z_0) = 0$, and vice versa.

Since both $B(z)$ and $S(z)$ are Blaschke products, we get

$$B(z) = \frac{B(i)}{S(i)}S(z). \tag{10.5}$$

To complete the proof it remains to observe that

$$\frac{B(i)}{S(i)} = \frac{B(i)}{\varkappa} = \frac{e^{-\ell} + k\Theta}{\Theta + e^{-\ell}k} \cdot \frac{k\Theta e^{-\ell} + 1}{k\Theta + e^{-\ell}} = \frac{k\Theta e^{-\ell} + 1}{\Theta + e^{-\ell}k}. \tag{10.6}$$

$\square$

Next, recall that by Theorem 5.7 the self-adjoint magnetic Hamiltonian D_Θ, $|\Theta| = 1$, on the metric graph

$$\mathbb{Y} = (-\infty, 0) \sqcup (0, \infty) \sqcup (0, \ell)$$

in Case (iii) referred to in Theorem 5.7 dilates the maximal dissipative differentiation operator $\widehat{d}_\Theta$.

Define the transmission coefficient $t(\lambda)$ in the scattering problem on the graph $\overline{\mathbb{Y}}$ (obtained from $\mathbb{Y}$ by identifying the right vertex of the edge $[0, \ell]$ with the vertex at the origin) as the amplitude of the generalized eigenfunction of the Hamiltonian D_Θ, the solution f to the equation

$$i\frac{d}{dx}f = \lambda f \quad \text{on} \quad \mathbb{Y} = (-\infty, 0) \sqcup (0, \infty) \sqcup (0, \ell) \tag{10.7}$$

that coincides with $e^{-i\lambda x}$ on the incoming edge $(-\infty, 0)$ of the graph $\mathbb{Y}$, equals $t(\lambda)e^{-i\lambda x}$ on the outgoing edge $(0, \infty)$, and $f = (f_\infty, f_\ell)$ satisfies the boundary conditions (5.3),

$$\begin{pmatrix} f_\infty(0+) \\ f_\ell(0) \end{pmatrix} = \begin{pmatrix} k & \sqrt{1 - k^2}\Theta \\ \sqrt{1 - k^2} & -k\Theta \end{pmatrix} \begin{pmatrix} f_\infty(0-) \\ f_\ell(\ell) \end{pmatrix}. \tag{10.8}$$

The analytic counterpart of the dilation Theorem (5.7) is as follows.

Theorem 10.2. *The transmission coefficient $t(\lambda)$ in the scattering problem* (10.7), (10.8) *has the form*

$$t(\lambda) = \frac{\Theta + e^{i\ell\lambda}k}{e^{i\ell\lambda} + k\Theta}, \quad \lambda \in \mathbb{R}. \tag{10.9}$$

Proof. Let f be the solution to the scattering problem (10.7).
   We have

$$f_\infty(\lambda, x) = e^{-i\lambda x}\chi_{(-\infty, 0)}(x) + t(\lambda)e^{-i\lambda x}\chi_{(0, \infty)}(x), \quad x \in \mathbb{R},$$

and

$$f_\ell(\lambda, x) = a(\lambda)e^{-i\lambda x}, \quad x \in (0, \ell).$$

From (5.3) it follows that

$$\begin{pmatrix} t(\lambda) \\ a(\lambda) \end{pmatrix} = \begin{pmatrix} k & \sqrt{1-k^2}\Theta \\ \sqrt{1-k^2} & -k\Theta \end{pmatrix} \begin{pmatrix} 1 \\ a(\lambda)e^{-i\lambda\ell} \end{pmatrix}.$$

Solving for $a(\lambda)$ we get that

$$a(\lambda) = \frac{\sqrt{1-k^2}}{1+k\Theta e^{-i\lambda\ell}}$$

and hence

$$t(\lambda) = k + \sqrt{1-k^2}\Theta e^{-i\lambda\ell}a(k) = k + \sqrt{1-k^2}\Theta e^{-i\lambda\ell}\frac{\sqrt{1-k^2}}{1+k\Theta e^{-i\lambda\ell}}$$

$$= \frac{\Theta + e^{i\ell\lambda}k}{e^{i\ell\lambda} + k\Theta}, \quad \lambda \in \mathbb{R}.$$

$\square$

Remark 10.3. We observe that if one sets $\Theta = 1$ in (10.9), then

$$t(\lambda) = \frac{1}{s_{\ell'}\left(\frac{\ell}{\ell'}\lambda\right)},$$

where $\ell' = \log\frac{1}{k}$ and

$$s_{\ell'}(z) = \frac{e^{iz\ell'} - e^{-\ell'}}{e^{-\ell'}e^{iz\ell'} - 1}$$

is the Livšic function associated with the pair $(\dot{D}, D)$ on the metric graph

$$\mathbb{Y} = (0, \ell')$$

in Case (ii) referred to in Lemma 6.2.

Corollary 10.4. *Let $t(\lambda)$ be the transmission coefficient in the scattering problem* (10.7), (10.8).
Then,

$$t(\lambda) = \frac{\Theta + e^{-\ell}k}{ke^{-\ell}\Theta + 1} \cdot S^{-1}(\lambda + i0), \quad \lambda \in \mathbb{R}, \qquad (10.10)$$

where $S(z)$ is the characteristic function of the triple $(\dot{d}, \widehat{d}_\Theta, d)$ in Case (ii).
In particular, the poles of the analytic (meromorphic) continuation of the transmission coefficient $t(\lambda)$ to the upper half-plane coincide with the eigenvalues of the dissipative operator $\widehat{d}_\Theta$.

Proof. Representation (10.10) follows from (10.1) and (10.9). Since $S(z)$ is analytic in $\mathbb{C}_+$, the transmission coefficient $t(\lambda)$ can be meromorphically continued on the whole complex plane. The (simple) poles of this continuation are located at the zeroes of the characteristic function $S(z)$ which are the eigenvalues of the dissipative operator $\hat{d}_\Theta$. ☐

Remark 10.5. The exceptional case of the triple $(\dot{d}, \hat{d}, d)$, where $\hat{d}$ is the differentiation operator on

$$\mathrm{Dom}(\hat{d}) = \{f_\ell \in W_2^1((0,\ell)) \,|\, f_\ell(0) = 0\} \tag{10.11}$$

deserves a special discussion. Notice that $\hat{d}$ is the only one dissipative quasi-selfadjoint extension of the symmetric differentiation $\dot{d}$ which is not in the family (5.21) with $k \neq 0$ (cf. Frostman's observation of general character: if S is an inner function, then $\frac{S-\varkappa}{1-\overline{\varkappa}S}$ is a pure Blaschke product for almost all $\varkappa \in \mathbb{D}$).

By Lemma 9.1, the characteristic function of the triple

$$S_{(\dot{d},\hat{d},d)} = e^{i\ell z}, \quad z \in \mathbb{C}_+,$$

is a singular inner function (see Remark 9.4).

On the other hand, the transmission coefficient of the self-adjoint dilation D of the dissipative operator $\hat{d}$ on the one-cycle graph can be evaluated by solving the equation

$$i\frac{d}{dx}f = \lambda f$$

on the metric graph $\mathbb{Y}$ with boundary conditions (5.3) with $k = 0$, and $\Theta = 1$,

$$\begin{pmatrix} f_\infty(0+) \\ f_\ell(0) \end{pmatrix} = \begin{pmatrix} 0 & 1 \\ 1 & 0 \end{pmatrix} \begin{pmatrix} f_\infty(0-) \\ f_\ell(\ell) \end{pmatrix}$$

to get an analog of (10.10). In this case,

$$t(\lambda) = e^{-i\ell z} = S_{(\dot{d},\hat{d},d)}^{-1}(\lambda), \quad \lambda \in \mathbb{R}.$$

Chapter 11

UNIQUENESS RESULTS

So far we were interested in characterizing solutions to the commutation relations (1.3) under the assumption that the unitary group U_t is given. Our next goal is to show that if the symmetric operator $\dot{A}$ (from Hypothesis 3.1) is a prime symmetric operator, then the commutation relations (1.3) uniquely determine the group U_t up to a character $t \to e^{it\mu}$ (with the only one exception).

We start with a preliminary observation that a prime symmetric operator with deficiency indices $(1,1)$ has a rather poor set of symmetries.

Lemma 11.1. *Suppose that $\dot{A}$ is a symmetric operator with deficiency indices $(1,1)$ and $\mathcal{U}$ is a unitary operator. Assume that the operator $\mathcal{U}$ commutes with $\dot{A}$ in the sense that*

$$\mathcal{U}(\mathrm{Dom}(\dot{A})) = \mathrm{Dom}(\dot{A}) \tag{11.1}$$

and

$$\dot{A}\mathcal{U}f = \mathcal{U}\dot{A}f \quad \text{for all } f \in \mathrm{Dom}(\dot{A}).$$

Then the subspaces

$$\mathcal{H}_\pm = \overline{\mathrm{span}_{z \in \mathbb{C}_\pm} \mathrm{Ker}((\dot{A})^* - zI)}$$

are invariant for $\mathcal{U}$. Moreover, the corresponding restrictions of $\mathcal{U}$ onto those subspaces are multiples of the identity. That is,

$$\mathcal{U}|_{\mathcal{H}_\pm} = \Theta_\pm I_{\mathcal{H}_\pm} \quad \text{for some } |\Theta_\pm| = 1. \tag{11.2}$$

Proof. Suppose that $f \in \mathrm{Dom}((\dot{A})^*)$. Then

$$(\dot{A}g, \mathcal{U}f) = (\mathcal{U}^* \dot{A}g, f) = (\mathcal{U}^* \dot{A} \mathcal{U} \mathcal{U}^* g, f) \quad \text{for all } g \in \mathrm{Dom}(\dot{A}).$$

From (11.1) it follows that $\mathcal{U}^* g \in \mathrm{Dom}(\dot{A})$, $g \in \mathrm{Dom}(\dot{A})$. Therefore,

$$(\mathcal{U}^* \dot{A} \mathcal{U} \mathcal{U}^* g, f) = (\dot{A} \mathcal{U}^* g, f) = (g, \mathcal{U}(\dot{A})^* f).$$

That is,

$$(\dot{A}g, \mathcal{U}f) = (g, \mathcal{U}(\dot{A})^* f) \quad \text{for all } g \in \mathrm{Dom}(\dot{A}),$$

which means that

$$\mathcal{U}(\mathrm{Dom}((\dot{A})^*)) \subset \mathrm{Dom}((\dot{A})^*)$$

and

$$(\dot{A})^* \mathcal{U} = \mathcal{U}(\dot{A})^* \quad \text{on } \mathrm{Dom}((\dot{A})^*). \tag{11.3}$$

Since (11.3) holds, the deficiency subspace

$$N_z = \mathrm{Ker}((\dot{A})^* - zI), \quad z \in \mathbb{C}_+,$$

is an eigensubspace of $\mathcal{U}$. Therefore, the subspace

$$\mathcal{H}_+ = \overline{\mathrm{span}_{z \in \mathbb{C}_+} \mathrm{Ker}((\dot{A})^* - zI)}$$

is invariant for $\mathcal{U}$.

Next we claim that the deficiency subspaces N_z and N_ζ are not orthogonal to each other for $z, \zeta \in \mathbb{C}_+$.

Indeed, let A be a self-adjoint extension of $\dot{A}$. Suppose that $g_+ \in N_i$ with $g_+ \neq 0$. Then the element $g_z = (A - iI)(A - zI)^{-1} g_+$ generates the subspace N_z, $\mathrm{Im}(z) \neq 0$. Therefore,

$$
\begin{aligned}
(g_z, g_\zeta) &= \left((A - iI)(A - zI)^{-1} g_+, (A - iI)(A - \zeta I)^{-1} g_+ \right) \\
&= \int_{\mathbb{R}} \frac{\lambda^2 + 1}{(\lambda - z)(\lambda - \overline{\zeta})} d\mu(\lambda) \\
&= \frac{1}{z - \overline{\zeta}} \int_{\mathbb{R}} \left(\frac{1}{\lambda - z} - \frac{1}{\lambda - \overline{\zeta}} \right) (\lambda^2 + 1) d\mu(\lambda),
\end{aligned}
$$

where $\mu(d\lambda)$ is the spectral measure of the element g_+ associated with the self-adjoint operator A. That is,

$$d\mu(\lambda) = (dE(\lambda)g_+, g_+),$$

where $E(\lambda)$ is the resolution of identity for the self-adjoint operator A,

$$A = \int_{\mathbb{R}} \lambda \, dE(\lambda).$$

Clearly,

$$\operatorname{Im}\left(\frac{1}{\lambda - z} - \frac{1}{\lambda - \bar{\zeta}}\right) > 0, \quad \text{whenever } z, \zeta \in \mathbb{C}_+,$$

and therefore

$$(g_z, g_\zeta) \neq 0, \quad z, \zeta \in \mathbb{C}_+,$$

which proves the claim.

Finally, since $\operatorname{Ker}((\dot{A})^* - zI)$ and $\operatorname{Ker}((\dot{A})^* - \zeta I)$ for $z, \zeta \in \mathbb{C}_+$ are not orthogonal to each other, the restrictions of $\mathcal{U}$ onto these subspaces have the same eigenvalues, proving that the restriction of $\mathcal{U}$ onto $\mathcal{H}_+$ is a multiple of the identity.

The same reasoning shows that the restriction of $\mathcal{U}$ onto its invariant subspace

$$\mathcal{H}_- = \overline{\operatorname{span}_{z \in \mathbb{C}_-} \operatorname{Ker}((\dot{A})^* - zI)}$$

is also a (possibly different) multiple of the identity as well.

The proof is complete. $\qquad\qquad\qquad\qquad\qquad\qquad\qquad\qquad\qquad\square$

Lemma 11.2. *Suppose that $\dot{A}$ is a symmetric operator with deficiency indices $(1,1)$. Assume that U_t and V_t are strongly continuous unitary groups such that the commutation relations*

$$U_t^* \dot{A} U_t = V_t^* \dot{A} V_t = \dot{A} + tI \quad \text{on } \operatorname{Dom}(\dot{A})$$

hold.

Then the subspaces

$$\mathcal{H}_\pm = \overline{\operatorname{span}_{z \in \mathbb{C}_\pm} \operatorname{Ker}((\dot{A})^* - zI)}$$

reduce the groups U_t and V_t and

$$V_t|_{\mathcal{H}_\pm} = e^{it\mu_\pm} U_t|_{\mathcal{H}_\pm} \quad \text{for some } \mu_\pm \in \mathbb{R}.$$

Proof. In a similar way as in the proof of Lemma 11.1, one observes that the commutation relations

$$U_t^*(\dot{A})^* U_t = V_t^*(\dot{A})^* V_t = (\dot{A})^* + tI \quad \text{on Dom}((\dot{A})^*)$$

for the adjoint operator $(\dot{A})^*$ hold.

Since obviously

$$U_t(\text{Ker}((\dot{A})^* - zI)) = V_t(\text{Ker}((\dot{A})^* - zI)) = (\text{Ker}((\dot{A})^* - (z - t)I)), \quad t \in \mathbb{R},$$

the subspaces $\mathcal{H}_\pm$ are invariant for both U_t and V_t for all t, and therefore $\mathcal{H}_\pm$ reduce the groups U_t and V_t. Since the unitary operator $\mathcal{U}_t = U_t^* V_t$ commutes with $\dot{A}$ on Dom($\dot{A}$), by Lemma 11.1 one gets that

$$U_t^* V_t|_{\mathcal{H}_\pm} = \mathcal{U}_t|_{\mathcal{H}_\pm} = e^{i\alpha_\pm(t)} I_{\mathcal{H}_\pm}, \quad t \in \mathbb{R}, \tag{11.4}$$

for some continuous real-valued functions $\alpha_\pm(t)$ (the continuity of the function $\alpha(t)$ follows from the hypothesis that the groups U_t and V_t are strongly continuous). That is,

$$V_t = e^{i\alpha_\pm(t)} U_t \quad \text{on } \mathcal{H}_\pm.$$

Since U_t and V_t are one-parameter groups, it follows that the functional equation

$$\alpha_\pm(t + s) = \alpha_\pm(t) + \alpha_\pm(s)$$

holds and hence, due to the continuity of $\alpha_\pm$, we conclude that

$$\alpha_\pm(t) = \mu_\pm t,$$

for some $\mu_\pm \in \mathbb{R}$, which combined with (11.4) completes the proof. $\square$

Theorem 11.3. *Suppose that $\widehat{A}$ is a maximal dissipative extension of a prime symmetric operator $\dot{A}$ with deficiency indices $(1, 1)$.*

Assume that U_t and V_t are strongly continuous unitary groups such that the commutation relations

$$U_t^* \widehat{A} U_t = V_t^* \widehat{A} V_t = \widehat{A} + tI \quad on \text{ Dom}(\widehat{A}) \tag{11.5}$$

hold.

If $\widehat{A}$ is self-adjoint assume, in addition, that

$$U_t^* \dot{A} U_t = V_t^* \dot{A} V_t = \dot{A} + tI \quad on \text{ Dom}(\dot{A}). \tag{11.6}$$

If $\widehat{A}$ has a regular point in the upper half-plane, in particular, if $\widehat{A}$ is self-adjoint, then

$$V_t = e^{it\mu} U_t \quad for \ some \ \mu \in \mathbb{R}. \tag{11.7}$$

Moreover, in the exceptional case when the spectrum of $\widehat{A}$ fills in the whole upper half-plane, the subspaces

$$\mathcal{H}_\pm = \overline{\mathrm{span}_{z \in \mathbb{C}_\pm} \mathrm{Ker}((\dot{A})^* - zI)}$$

are orthogonal, reduce the groups U_t and V_t and

$$V_t|_{\mathcal{H}_\pm} = e^{it\mu_\pm} U_t|_{\mathcal{H}_\pm} \quad \text{for some } \mu_\pm \in \mathbb{R}.$$

Proof. Let z be a regular point of $\widehat{A}$ in the upper half-plane. As it has been explained in the proof of Theorem 3.5, if $\widehat{A}$ is not self-adjoint, then (11.6) holds automatically. Therefore, Lemma 11.2 is applicable and hence

$$\mathcal{U}_t g = e^{i\mu_- t} g \quad \text{for some } \mu_- \in \mathbb{R},$$

where $\mathcal{U}_t = U_t^* V_t$ and $0 \neq g \in \mathrm{Ker}((\dot{A})^* + iI) \subset \mathcal{H}_-$.

Set

$$f = (\widehat{A} + iI)(\widehat{A} - zI)^{-1} g.$$

Since $f \in \mathrm{Ker}((\dot{A})^* - zI) \subset \mathcal{H}_+$, by Lemma 11.2,

$$\mathcal{U}_t f = e^{i\mu_+ t} f \quad \text{for some } \mu_+ \in \mathbb{R}.$$

On the other hand, since the unitary operator $\mathcal{U}_t$ commutes with $\widehat{A}$, we have

$$\mathcal{U}_t f = (\widehat{A} + iI)(\widehat{A} - zI)^{-1} \mathcal{U}_t g = (\widehat{A} + iI)(\widehat{A} - zI)^{-1} e^{i\mu_- t} g = e^{i\mu_- t} f.$$

Therefore, $\mu_+ = \mu_-$.

Finally, taking into account that $\dot{A}$ is a prime operator, it follows that the subspaces $\mathcal{H}_\pm$ span the whole Hilbert space $\mathcal{H}$, and the claim follows.

In view of Lemma 11.2, to prove the last assertion it remains to show that $\mathcal{H}_\pm$ are mutually orthogonal whenever the spectrum of $\widehat{A}$ fills in the upper half-plane.

Since (11.5) holds and the dissipative operator $\widehat{A}$ has no regular points in the upper half-plane, one can apply Theorem 3.5 to conclude that for any self-adjoint extension of $\dot{A}$ the characteristic function of the triple $(\dot{A}, \widehat{A}, A)$ is identically zero. By Lemma 9.1, the characteristic function of the triple $(\dot{D}, \widehat{D}_I(0), D)$ on the metric graph $\mathbb{Y}$ in Case (i) with quantum gate coefficient $k = 0$ also vanishes identically in the upper-half-plane. The operators $\dot{A}$ and $\dot{D}$ are prime symmetric operators, therefore $\dot{A}$ is unitarily equivalent to $\dot{D}$ on the metric graph $\mathbb{Y}$ in Case (i), where $\widehat{D}$ is in the exceptional case, that is, its point spectrum fills in $\mathbb{C}_+$. In this case the subspaces $\mathcal{H}_\pm(\dot{D}) = \overline{\text{span}_{z \in \mathbb{C}_\pm} \text{Ker}((\dot{D})^* - zI)} = L^2(\mathbb{R}_\pm)$ for the operator $\dot{D}$ are orthogonal, so are the subspaces

$$\mathcal{H}_\pm(\dot{A}) = \overline{\text{span}_{z \in \mathbb{C}_\pm} \text{Ker}((\dot{A})^* - zI)}$$

for the symmetric operator $\dot{A}$. $\qquad\qquad\qquad\qquad\qquad\qquad\qquad$ $\square$

Chapter 12

DISSIPATIVE SOLUTIONS TO THE CCR

Now we are prepared to get a complete classification (up to unitary equivalence) of the simplest non-self-adjoint maximal dissipative solutions $\widehat{A}$ to the commutation relations (1.4).

More generally, we have the following result.

Theorem 12.1. *Assume Hypothesis 3.1. Suppose, in addition, that $\dot{A}$ is a prime operator and A its self-adjoint extension. Suppose that $\widehat{A}$ is a maximal dissipative extension of $\dot{A}$ such that*

$$U_t^* \widehat{A} U_t = \widehat{A} + tI \quad on \ \mathrm{Dom}(\widehat{A}).$$

(i) *If $\widehat{A}$ is self-adjoint, then there exists a unique Θ, $|\Theta| = 1$, such that the triple $(\dot{A}, \widehat{A}, A)$ is mutually unitarily equivalent to the triple $(\dot{D}, D, D_\Theta)$ on the metric graph $\mathbb{Y}$ in Case (i) and therefore Θ is a unitary invariant of $(\dot{A}, \widehat{A}, A)$.*

(ii) *If $\widehat{A}$ is not self-adjoint, then the triple $(\dot{A}, \widehat{A}, A)$ is mutually unitarily equivalent to the model triple $(\dot{D}, \widehat{D}, D_\Theta)$ in one of the Cases (i)–(iii) for some $|\Theta| = 1$. In addition, if $\widehat{A}$ has at least one regular point in the upper half-plane, then the parameter Θ is uniquely determined by the triple $(\dot{A}, \widehat{A}, A)$ and therefore Θ is a unitary invariant of $(\dot{A}, \widehat{A}, A)$ in this case. That is, if some triples $(\dot{A}, \widehat{A}, A)$ and $(\dot{A}', \widehat{A}', A')$ are mutually unitarily equivalent, then the corresponding parameters Θ and Θ' coincide.*

Proof. (i) If $\widehat{A}$ is self-adjoint, we argue as follows. By Theorem 3.3, the Weyl-Titchmarsh function of $(\dot{A}, \widehat{A})$ coincides with i in the upper half-plane. Therefore, since $\dot{A}$ is a prime operator, the pair $(\dot{A}, \widehat{A})$ is mutually unitarily equivalent to the pair $(\dot{D}, D)$ on the metric graph $\mathbb{Y}$ in Case (i). Since A is a self-adjoint extension of $\dot{A}$, the triple $(\dot{A}, \widehat{A}, A)$ is mutually unitarily equivalent to the triple $(\dot{D}, D, D_\Theta)$ for some $|\Theta| = 1$, which proves the existence of such a Θ.

To establish the uniqueness, suppose that the triples $(\dot{D}, D, D_{\Theta_1})$ and $(\dot{D}, D, D_{\Theta_2})$ in Case (i) (recall that D is self-adjoint here) are mutually unitarily equivalent.

In particular, there exists a unitary operator U such that

$$U(\text{Dom}(\dot{D}) = \text{Dom}(\dot{D})), \quad U(\text{Dom}(D)) = \text{Dom}(D),$$

$$U\dot{D}f = \dot{D}Uf \quad \text{for all } f \in \text{Dom}(\dot{D}),$$

$$UDf = DUf \quad \text{for all } f \in \text{Dom}(D)$$

and

$$U^* D_{\Theta_1} U = D_{\Theta_2}.$$

By Lemma 11.2, the subspaces $L^2(\mathbb{R}_\pm)$ are eigensubspaces for the unitary operator U, and since $U(\text{Dom}(D)) = \text{Dom}(D)$, the corresponding eigenvalues of U coincide. Therefore, U is necessarily a (unimodular) multiple of the identity and hence

$$D_{\Theta_2} = U^* D_{\Theta_1} U = D_{\Theta_1}$$

so that

$$\Theta_1 = \Theta_2.$$

(ii) Suppose that $\widehat{A}$ is not self-adjoint. By Theorem 3.5, the characteristic function of the triple admits the representation (3.7). Combining Theorem 3.5 and Lemma 9.1 one concludes that there exists a (possibly different) self-adjoint extension A' of $\dot{A}$ such that the triples $(\dot{A}, \widehat{A}, A')$ and $(\dot{D}, \widehat{D}, D)$ are mutually unitarily equivalent.

In particular, $\mathcal{U}^{-1}\dot{A}\mathcal{U} = \dot{D}$ and $\mathcal{U}^{-1}A'\mathcal{U} = D$ for some unitary operator. Hence $\mathcal{U}^{-1}A\mathcal{U}$ is a self-adjoint extension of $\dot{D}$ and hence $\mathcal{U}^{-1}A\mathcal{U} = D_\Theta$ for some Θ. It is the unitary transformation $\mathcal{U}$ that establishes the required mutual unitary equivalence of the triples.

To prove the last assertion, one observes that since $\widehat{A}$ has at least one regular point in the upper half-plane, the characteristic function of the triple is not identically zero. In particular, if A' is another reference operator, then the triples $(\dot{A}, \widehat{A}, A)$ and $(\dot{A}, \widehat{A}, A')$ are mutually unitarily equivalent if and only if A' and A coincide by Lemma E.1 in Appendix E.

Therefore, taking into account that the triple $(\dot{A}, \widehat{A}, A)$ is mutually unitarily equivalent to $(\dot{D}, \widehat{D}, D_\Theta)$ for some $|\Theta| = 1$, one concludes that in this case the unimodular parameter Θ is uniquely determined by the triple $(\dot{A}, \widehat{A}, A)$. □

Remark 12.2. If $\widehat{A}$ has no regular points in the upper half-plane, then $(\dot{A}, \widehat{A}, A')$ is mutually unitarily equivalent to $(\dot{D}, \widehat{D}, D_\Theta)$ for some (and therefore for all) Θ, $|\Theta| = 1$, where $\widehat{D}$ is in the exceptional case (Case (i) with $k = 0$) (see Corollary 9.5). Therefore, in this exceptional case, the parameter Θ is not determined uniquely by the triple $(\dot{A}, \widehat{A}, A')$.

Remark 12.3. On account of the remarks that we made in Chapter 8, structure Theorem 8.3 combined with Theorem 12.1 provides the following intrinsic characterization of all symmetric operators satisfying Hypothesis 3.1 thus giving a *complete solution of the Jørgensen-Muhly problem* for symmetric operators in the case of deficiency indices $(1, 1)$ (see Problem (I) b) in the Introduction):
either

i) $\dot{A}$ admits a (dissipative) quasi-selfadjoint extension with the point spectrum filling in $\mathbb{C}_+$, equivalently, $\dot{A}$ admits a pair of distinct quasi-selfadjoint extensions that are unitarily equivalent,
   or,

ii) $\dot{A}$ admits a quasi-selfadjoint extension with no spectrum,
   or, finally,

iii) $\dot{A}$ is the symmetric part of an operator coupling of a dissipative extension of the symmetric operator $\dot{A}$ without point spectrum in case i) and the dissipative extension of $\dot{A}$ with no spectrum in case ii) (see Remark 3.6 for the definition of the symmetric part of a dissipative operator in connection with Hypothesis 3.1).

Chapter 13

MAIN RESULTS

In this chapter we provide the complete classification of the simplest solutions to the restricted Weyl commutation relations

$$V_s U_t = e^{ist} U_t V_s, \quad t \in \mathbb{R}, \quad s \geq 0,$$

for a strongly continuous group of unitary operators U_t and a strongly continuous semi-group of contractions V_s in a separable Hilbert space $\mathcal{H}$.

Hypothesis 13.1. Let $(-\infty, \infty) \ni t \to U_t = e^{iBt}$ be a strongly continuous group of unitary operators and $[0, \infty) \ni s \to V_s = e^{i\widehat{A}s}$ a semi-group of contractions in a separable Hilbert space $\mathcal{H}$. Suppose that the restricted Weyl commutation relations

$$V_s U_t = e^{ist} U_t V_s, \quad t \in \mathbb{R}, \quad s \geq 0, \tag{13.1}$$

hold.

We remark that in the light of Corollary 6.4 one could have started from the much weaker Hypothesis 3.1.

The following two results characterize the simplest solutions to the restricted Weyl commutation relations. We start with the case where V_s is a semi-group of isometries.

Theorem 13.2. *Assume Hypothesis* 13.1. *Suppose, in addition, that the generator $\widehat{A}$ of the semi-group V_s is a prime symmetric operator with deficiency indices* $(0, 1)$.

Then there exists a unique metric graph $\mathbb{Y} = (\mu, \infty)$, $\mu \in \mathbb{R}$, such that the pair $(\widehat{A}, B)$ is mutually unitarily equivalent to the pair $(\widehat{\mathcal{P}}, \mathcal{Q})$, where

$\widehat{\mathcal{P}} = i\frac{d}{dx}$ *is the differentiation operator in* $L^2(\mathbb{Y}) = L^2((\mu, \infty))$ *on*

$$\text{Dom}(\widehat{\mathcal{P}}) = \{f \in W_2^1((\mu, \infty)) \mid f(\mu) = 0\}$$

and $\mathcal{Q}$ *is the operator of multiplication by independent variable in* $L^2(\mathbb{Y})$.

Proof. Since $\widehat{A}$ is a generator of a semi-group, $\widehat{A}$ is a closed operator. By the Stone-von Neumann uniqueness result (see, e.g., [3, Theorem 2, Ch. VIII, Sec. 104]) there exists a isometric map $\mathcal{U}$ from $\mathcal{H}$ onto $L^2((0, \infty))$ such that $\mathcal{U}\widehat{A}\mathcal{U}^{-1}$ coincides with the differentiation operator in $L^2((0, \infty))$ with the Dirichlet boundary condition at the origin.

   Lemma 11.2 and Lemma B.5 in Appendix B show that there exists a μ such that the pair $(\widehat{A}, B)$ is mutually unitarily equivalent to the pair $(\mathcal{P}_0, \mathcal{Q}_0 + \mu I)$, where

$$\mathcal{P}_0 = i\frac{d}{dx}$$

is the differentiation operator in $L^2((0, \infty))$ on

$$\text{Dom}(\mathcal{P}_0) = \{f \in W_2^1((0, \infty)) \mid f(0) = 0\},$$

and $\mathcal{Q}_0$ is the operator of multiplication by independent variable in $L^2((0, \infty))$. Clearly the pairs $(\mathcal{P}_0, \mathcal{Q}_0 + \mu I)$ and $(\widehat{\mathcal{P}}, \mathcal{Q})$ in the Hilbert spaces $L^2((0, \infty))$ and $L^2((\mu, \infty))$, respectively, are mutually unitarily equivalent. By construction, the spectrum of the generator B coincides with the semi-axis $[\mu, \infty)$. Therefore, μ is a unitary invariant which is uniquely determined by the pair $(\widehat{A}, B)$. In particular, the graph $\mathbb{Y} = (\mu, \infty)$ is also uniquely determined by the pair $(\widehat{A}, B)$. $\qquad\square$

   In a completely analogous way one proves the following result.

Theorem 13.3. *Assume Hypothesis 13.1. Suppose, in addition, that the generator* $\widehat{A}$ *of the semi-group* V_s *is a maximal dissipative extension of a prime symmetric operator with deficiency indices* $(1, 0)$.

   Then there exists a unique metric graph $\mathbb{Y} = (-\infty, \nu)$, $\nu \in \mathbb{R}$, *such that the pair* $(\widehat{A}, B)$ *is mutually unitarily equivalent to the pair* $(\widehat{\mathcal{P}}, \mathcal{Q})$ *where* $\widehat{\mathcal{P}} = i\frac{d}{dx}$ *is the differentiation operator in* $L^2(\mathbb{Y}) = L^2((-\infty, \nu))$ *on*

$$\text{Dom}(\widehat{\mathcal{P}}) = W_2^1((-\infty, \nu))$$

and $\mathcal{Q}$ *is the operator of multiplication by independent variable in* $L^2(\mathbb{Y})$.

Remark 13.4. We remark that if $\widehat{A}$ is a prime symmetric operator with deficiency indices $(0, 1)$ or $(1, 0)$ in a Hilbert space $\mathcal{H}$, then $\mathcal{H}$ is necessarily separable.

Remark 13.5. Theorems 13.2 and 13.3 are dual to each other: if $(\widehat{A}, B)$ satisfies the hypotheses of Theorem 13.2, then $(-\widehat{A}^*, -B)$ satisfies the hypotheses of Theorem 13.3, however the pairs $(\widehat{A}, B)$ and $(-\widehat{A}^*, -B)$ are not mutually unitarily equivalent. In this case the self-adjoint operator B is semi-bounded from below but $-B$ is semi-bounded from above. Moreover, the generator $\widehat{A}$ has no point spectrum while the point spectrum of $(-\widehat{A})^*$ fills in the entire open upper half-plane.

Next we treat the case where the dissipative generator of the semi-group V_s is a quasi-selfadjoint extension of a prime symmetric operator with deficiency indices $(1, 1)$.

To formulate the corresponding uniqueness result, we need some preparations.

Definition 13.6. Let $\mathbb{Y}$ be a metric graph in the following cases

$$\mathbb{Y} = \begin{cases} (-\infty, \nu) \sqcup (\mu, \infty) & \text{Case I}^* \quad (\nu \neq \mu) \\ (-\infty, \mu) \sqcup (\mu, \infty) & \text{Case I} \\ (\mu, \nu) & \text{Case II} \quad (\mu < \nu) \\ (-\infty, \mu) \sqcup (\mu, \infty) \sqcup (\mu, \nu) & \text{Case III} \quad (\mu < \nu). \end{cases} \tag{13.2}$$

Given a real number k, $0 \leq k < 1$, define

the *position* operator $\mathcal{Q}$ as the operator of multiplication by independent variable on the edges of the graph $\mathbb{Y}$

and

the *momentum* operator $\widehat{\mathcal{P}}$ as the differentiation operator $i\frac{d}{dx}$ on the edges of the graph $\mathbb{Y}$,

$$(\widehat{\mathcal{P}}f)(x) = i\frac{d}{dx}f(x) \quad \text{a. e. } x \in e \text{ on every edge } e \text{ of } \mathbb{Y} \tag{13.3}$$

on

$$\mathrm{Dom}(\widehat{\mathcal{P}}) \subsetneq \bigoplus_{e \subset \mathbb{Y}} W_2^1(e) \subsetneq L^2(\mathbb{Y}),$$

the space of locally absolutely continuous functions on the edges with the following vertex boundary conditions

$$\mathrm{Dom}(\widehat{\mathcal{P}}) = \begin{cases} \{f_- \oplus f_+ \in W_2^1((-\infty,\nu)) \oplus W_2^1((\mu,\infty)) \mid f_+(\mu+) = 0\}, \\ \quad \text{in Case I}^* \\ \{f_\infty \in W_2^1((-\infty,\mu)) \oplus W_2^1((\mu,\infty)) \mid f_\infty(\mu+) = kf_\infty(\mu-)\}, \\ \quad \text{in Case I} \\ \{f_\ell \in W_2^1((\mu,\nu)) \mid f_\ell(\mu+) = 0\}, \\ \quad \text{in Case II.} \end{cases}$$

Here, in Case I we require that $0 < k < 1$ and in Case II we assume that $\nu = \mu + \ell > \mu$.

In Case III, $\mathrm{Dom}(\widehat{\mathcal{P}})$ consists of the two-component vector-functions $f = (f_\infty, f_\ell)^T$,

$$f_\infty \oplus f_\ell \in \left(W_2^1((-\infty,\mu)) \oplus W_2^1((\mu,\infty)) \right) \oplus W_2^1((\mu,\nu))$$

that satisfy the "boundary conditions"

$$f_\infty(\mu+) = kf_\infty(\mu-) \quad \text{and} \quad f_\ell(\mu+) = \sqrt{1-k^2}f_\infty(\mu-) \quad \text{(Case III)}.$$

$$(13.4)$$

By definition the pair $(\widehat{\mathcal{P}}, \mathcal{Q})$ is said to be the dissipative *canonical pair* *with the quantum gate coefficient* k, $0 \le k < 1$ on the metric graph $\mathbb{Y}$. In Case III we always assume that $k > 0$ and formally set $k = 0$ whenever the graph $\mathbb{Y}$ is in Case I* or in Case II. We also call the triple $(\dot{\mathcal{P}}, \widehat{\mathcal{P}}, \mathcal{Q})$, where

$$\dot{\mathcal{P}} = \widehat{\mathcal{P}}|_{\mathrm{Dom}(\widehat{\mathcal{P}}) \cap \mathrm{Dom}(\widehat{\mathcal{P}}_*)},$$

the canonical dissipative triple on $\mathbb{Y}$ *with the quantum gate coefficient* k.

Remark 13.7. In Case I* the metric graph is "disconnected" whenever $\nu < \mu$, while if $\mu < \nu$, one may think that the edges of the graph eventually "overlap" over the finite interval $[\mu,\nu]$. Also, in Case III the boundary conditions (13.4) at the junction point μ of the graph $\mathbb{Y}$, the center of the graph, yield the quantum Kirchhoff rule

$$|f_\infty(\mu+)|^2 + |f_\ell(\mu+)|^2 = |f_\infty(\mu-)|^2.$$

Remark 13.8. It is easy to see that the spectrum of the position operator $\mathcal{Q}$ is given by

$$\text{spec}(\mathcal{Q}) = \begin{cases} (-\infty, \nu] \cup [\mu, \infty), & \text{in Case I}^* \\ (-\infty, \infty), & \text{in Case I} \\ [\mu, \nu], & \text{in Case II} \\ (-\infty, \infty), & \text{in Case III} \end{cases} \tag{13.5}$$

From (13.5) it follows that if $\mathbb{Y}$ is in Case I* with $\nu > \mu$ or in Case III, then the spectrum of the position operator $\mathcal{Q}$ has multiplicity 2 on the finite interval $[\mu, \nu]$.

In Case I and II the position operator Q has simple Lebesgue spectrum filling in the whole real axis $(-\infty, \infty)$ and the finite interval $[\mu, \nu]$ respectively.

We also notice that the spectrum of the dissipative momentum operator $\widehat{\mathcal{P}}$ is

$$\text{spec}(\widehat{\mathcal{P}}) = \begin{cases} \mathbb{C}_+ \cup (-\infty, \infty), & \text{in Case I}^* \text{ and Case I with } k = 0 \\ (-\infty, \infty), & \text{in Case I with } k > 0 \\ \emptyset, & \text{in Case II} \\ (-\infty, \infty), & \text{in Case III} \end{cases} \tag{13.6}$$

Notice that Case I* and Case I with $k = 0$ are exceptional in the sense that any point in the (open) upper half-plane is an eigenvalue of $\widehat{\mathcal{P}}$.

If a metric graph $\mathbb{Y}$ is in Case I, we also introduce the concept of *the Weyl canonical triple* on $\mathbb{Y}$.

Definition 13.9. Let $\mathbb{Y}$ be a metric graph in Case I, that is,

$$\mathbb{Y} = (-\infty, \mu) \sqcup (\mu, \infty) \quad \text{for some } \mu \in \mathbb{R}.$$

Let $\mathcal{P} = i\frac{d}{dx}$ be the self-adjoint differentiation operator on

$$\text{Dom}(\mathcal{P}) = \{f \in W_2^1(\mathbb{Y}) \,|\, f(\mu - 0) = f(\mu + 0)\}, \tag{13.7}$$

$\dot{\mathcal{P}}$ its symmetric restriction on

$$\text{Dom}(\dot{\mathcal{P}}) = \{f \in W_2^1(\mathbb{Y}) \,|\, f(\mu - 0) = f(\mu + 0) = 0\},$$

and $\mathcal{Q}$ the position operator on $\mathbb{Y}$. We call $(\dot{\mathcal{P}}, \mathcal{P}, \mathcal{Q})$ *the Weyl canonical triple on* $\mathbb{Y}$ (centered at μ).

Remark 13.10. If $(\widehat{\mathcal{P}}, \mathcal{Q}) = (\widehat{\mathcal{P}}(k), \mathcal{Q})$ is the dissipative *canonical pair* on the metric graph $\mathbb{Y} = (-\infty, \mu] \cup [\mu, \infty)$ *with the quantum gate coefficient* k, then

$$\underset{k \to 1}{\text{s-lim}}\, \widehat{\mathcal{P}}(k) = \mathcal{P},$$

where $\mathcal{P}$ is the self-adjoint differentiation operator defined on (13.7) and the limit is taken in the strong resolvent sense. Therefore, the Weyl canonical triple on the metric graph $\mathbb{Y}$ can be considered the limiting case of the dissipative triple $(\dot{\mathcal{P}}, \widehat{\mathcal{P}}, \mathcal{Q}) = (\dot{\mathcal{P}}(k), \widehat{\mathcal{P}}(k), \mathcal{Q})$ with the quantum gate coefficient k as $k \to 1$.

Our first auxiliary result is that the pair $(\mathbb{Y}, k)$ is a unitary invariant of a dissipative canonical pair.

Lemma 13.11. *Suppose that the canonical pairs* $(\widehat{\mathcal{P}}(k), \mathcal{Q})$ *and* $(\widehat{\mathcal{P}}'(k'), \mathcal{Q}')$ *with the quantum gate coefficients* k *and* k' *on metric graphs* $\mathbb{Y}$ *and* $\mathbb{Y}'$, *respectively, are mutually unitarily equivalent. Then*

$$\mathbb{Y} = \mathbb{Y}' \quad and \quad k = k'.$$

Proof. As it has been explained in Remark 13.8, there are two options: either the position operator $\mathcal{Q}$ has simple spectrum or $\mathcal{Q}$ has spectrum of multiplicity 2 filling in a finite interval.

First, assume that the position operator $\mathcal{Q}$ has spectrum of multiplicity 2 supported by a finite interval $[\mu, \nu]$, $\nu > \mu$. So does $\mathcal{Q}'$. Therefore the graphs $\mathbb{Y}$ and $\mathbb{Y}'$ have the same vertices but may possibly be in different cases, in Case I* or in Case III only.

Suppose that $\mathbb{Y}$ is in Case I*and therefore $k = 0$. Then the point spectrum of the dissipative momentum operator $\widehat{\mathcal{P}} = \widehat{\mathcal{P}}(0)$ fills in the whole upper half-plane $\mathbb{C}_+$, so does the dissipative momentum operator $\widehat{\mathcal{P}}'$ since $\widehat{\mathcal{P}}$ and $\widehat{\mathcal{P}}'$ are unitarily equivalent. Therefore, $\mathbb{Y}'$ is in Case I* with $k' = 0$ as well. Analogously, if $\mathbb{Y}$ is in Case III, then $\mathbb{C}_+$ belongs to the resolvent set of $\widehat{\mathcal{P}}(k)$. Again, since $\widehat{\mathcal{P}}(k)$ and $\widehat{\mathcal{P}}'(k')$ are unitarily equivalent, $\mathbb{C}_+$ belongs to the resolvent set of $\widehat{\mathcal{P}}'(k')$ and then necessarily $\mathbb{Y}'$ is in Case III. Thus $\mathbb{Y}$ and $\mathbb{Y}'$ have the same vertices and are in the same cases. Therefore, $\mathbb{Y} = \mathbb{Y}'$.

It remains to treat the case where the multiplication operator $\mathcal{Q}$ has simple spectrum. There are two options: either both $\mathbb{Y}$ and $\mathbb{Y}'$ are in Case II, or both of them are in Case I.

If they are in Case II, the knowledge of the spectrum of $\mathcal{Q}$ ($\mathcal{Q}'$) uniquely determines the location of the vertices of the graph $\mathbb{Y}$ ($\mathbb{Y}'$) and the graph(s) itself.

If both $\mathbb{Y}$ and $\mathbb{Y}'$ are in Case I, we proceed as follows. Since the pairs $(\widehat{\mathcal{P}}(k), \mathcal{Q})$ and $(\widehat{\mathcal{P}}'(k'), \mathcal{Q}')$ are mutually unitarily equivalent and $\mathcal{Q} = \mathcal{Q}'$, there exists a unitary operator U commuting with the multiplication operator $\mathcal{Q}$ such that

$$\widehat{\mathcal{P}}'(k') = U^*\widehat{\mathcal{P}}(k)U.$$

Since $\mathcal{Q}$ has simple spectrum and the unitary operator U commutes with $\mathcal{Q}$, the operator U is the multiplication operator by a unimodular function u. We have

$$\mathrm{Dom}(\widehat{\mathcal{P}}(k)) = U(\mathrm{Dom}(\widehat{\mathcal{P}}'(k'))). \tag{13.8}$$

Suppose that the vertices μ and μ' of the graphs $\mathbb{Y}$ and $\mathbb{Y}'$ are different, that is, $\mu \neq \mu'$. From (13.8) it follows that the function $u(x)f(x)$ is a continuous function in a neighborhood of the point μ' for all $f \in \mathrm{Dom}(\widehat{\mathcal{P}}'(k'))$, so is the function $|f(x)| = |u(x)f(x)|$, which is incompatible with the boundary condition $f(\mu'+) = k'f(\mu'-)$ for all $f \in \mathrm{Dom}(\widehat{\mathcal{P}}'(k'))$, since $k' < 1$. Therefore, the vertices of the graphs $\mathbb{Y}$ and $\mathbb{Y}'$ coincide, $\mu = \mu'$, and hence $\mathbb{Y} = \mathbb{Y}'$.

To prove that $k = k'$, notice that if the metric graph $\mathbb{Y}$ and therefore $\mathbb{Y}'$ is in Cases I* or II, then $k = k' = 0$ by definition.

Suppose that $\mathbb{Y} = \mathbb{Y}'$ is in Cases I,

$$\mathbb{Y} = (-\infty, \mu) \sqcup (\mu, \infty) = \mathbb{Y}'.$$

In this case, the absolute values of the von Neumann parameters of $\widehat{\mathcal{P}}(k)$ and $\widehat{\mathcal{P}}'(k')$ (more precisely, of the corresponding triples) coincide with k and k', respectively. By the hypothesis, $\widehat{\mathcal{P}}(k)$ and $\widehat{\mathcal{P}}'(k')$ are unitarily equivalent. Therefore,

$$k = k',$$

since the absolute value of the von Neumann parameter is a unitary invariant of a dissipative operator by Remark 2.5.

Next, assume that $\mathbb{Y} = \mathbb{Y}'$ is in Case III,

$$\mathbb{Y} = (-\infty, \mu) \sqcup (\mu, \infty) \sqcup (\mu, \nu) = \mathbb{Y}' \quad (\mu < \nu).$$

By Theorem 9.3, the absolute values of the von Neumann parameters of $\widehat{\mathcal{P}}(k)$ and $\widehat{\mathcal{P}}'(k')$ are $ke^{-\ell}$ and $k'e^{-\ell}$, respectively (see the relation (2.8)), where

$$\ell = \nu - \mu.$$

Therefore, $k = k'$, since $\widehat{\mathcal{P}}(k)$ and $\widehat{\mathcal{P}}'(k')$ are unitarily equivalent by the hypothesis. □

Now we are ready to present the central result of the first part of the book.

Theorem 13.12. *Assume Hypothesis 13.1. Suppose, in addition, that the generator $\widehat{A}$ of the semi-group V_s is not self-adjoint and that the restriction*

$$\dot{A} = \widehat{A}\big|_{\mathrm{Dom}(\widehat{A}) \cap \mathrm{Dom}((\widehat{A})^*)}$$

is a prime symmetric operator with deficiency indices $(1, 1)$.

Then there exists a unique metric graph $\mathbb{Y}$ in one of the Cases I^*, I–III and a unique $k \in [0, 1)$ such that the pair $(\widehat{A}, B)$ (triple $(\dot{A}, \widehat{A}, B)$) is mutually unitarily equivalent to the canonical dissipative pair $(\widehat{\mathcal{P}}(k), \mathcal{Q})$ (triple $(\dot{\mathcal{P}}, \widehat{\mathcal{P}}(k), \mathcal{Q})$) on $\mathbb{Y}$, respectively.

Proof. By the hypothesis the restricted Weyl relations (13.1) hold. Therefore (see [27, 125])

$$U_t^* \widehat{A} U_t = \widehat{A} + tI \quad \text{on } \mathrm{Dom}(\widehat{A}), \quad t \in \mathbb{R}. \tag{13.9}$$

As in the proof of Theorem 3.5 one shows that symmetric operator $\dot{A}$ solves the commutation relations

$$U_t^* \dot{A} U_t = \dot{A} + tI \quad \text{on } \mathrm{Dom}(\dot{A}). \tag{13.10}$$

Therefore, the operator $\dot{A}$ satisfies Hypothesis 3.1. In this situation one can apply Theorem 12.1 (ii) to see that there is a metric graph $\mathbb{Y}_0$ in one of the Cases (i)–(iii) with the quantum gate coefficient k such that the dissipative operator $\widehat{A}$ is unitarily equivalent to the one of the following model dissipative differentiation operators:

(i) $\widehat{D} = \widehat{D}_I(k) = i\frac{d}{dx}$ on $\mathbb{Y}_0 = (-\infty, 0) \sqcup (0, \infty)$ with the boundary condition

$$f_\infty(0+) = k f_\infty(0-), \quad 0 \le k < 1; \tag{13.11}$$

or

(ii) $\widehat{D} = \widehat{D}_{II}(0, \ell) = i\frac{d}{dx}$ on $\mathbb{Y}_0 = (0, \ell)$ with the boundary condition

$$f_\ell(0) = 0; \tag{13.12}$$

or

(iii) $\widehat{D} = \widehat{D}_{III}(k, \ell) = i\frac{d}{dx}$ on $\mathbb{Y}_0 = (-\infty, 0) \sqcup (0, \infty) \sqcup (0, \ell)$ with the boundary conditions

$$\begin{pmatrix} f_\infty(0+) \\ f_\ell(0) \end{pmatrix} = \begin{pmatrix} k & 0 \\ \sqrt{1-k^2} & 0 \end{pmatrix} \begin{pmatrix} f_\infty(0-) \\ f_\ell(\ell) \end{pmatrix}, \quad 0 < k < 1. \tag{13.13}$$

That is, there exists a unitary map $\mathcal{W}$ from $\mathcal{H}$ onto $L^2(\mathbb{Y}_0)$ such that

$$\mathcal{W}\widehat{A}\mathcal{W}^{-1} = \widehat{D}. \tag{13.14}$$

In particular, from (13.9) it follows that

$$W_t^* \widehat{D} W_t = \widehat{D} + tI \quad \text{on } \mathrm{Dom}(\widehat{D}), \quad t \in \mathbb{R}, \tag{13.15}$$

where W_t is the unitary group on $L^2(\mathbb{Y})$ given by

$$W_t = \mathcal{W} U_t \mathcal{W}^{-1}, \quad t \in \mathbb{R}.$$

On the other hand,

$$e^{it\mathcal{Q}_0} \widehat{D} e^{-it\mathcal{Q}_0} = \widehat{D} + tI \quad \text{on } \mathrm{Dom}(\widehat{D}), \quad t \in \mathbb{R},$$

where $\mathcal{Q}_0$ is the operator of multiplication by independent variable on the graph $\mathbb{Y}_0$.

Applying Theorem 11.3 to the dissipative operator $\widehat{D}$, one obtains

$$W_t = \mathcal{W}e^{iBt}\mathcal{W}^{-1} = e^{-i\mu t}e^{i\mathcal{Q}_0 t} \quad \text{for some } \mu \in \mathbb{R}, \tag{13.16}$$

whenever $\widehat{D}$, and therefore $\widehat{A}$, has a regular point in the upper half-plane.

In this case, combining (13.14) and (13.16) one concludes that the pair $(\widehat{A}, B)$ is mutually unitarily equivalent to the pair $(\widehat{D}, \mathcal{Q} - \mu I)$ on the graph $\mathbb{Y}_0$ (with the quantum gate coefficient k). The pair $(\widehat{D}, \mathcal{Q} - \mu I)$ is in turn mutually unitarily equivalent to the canonical dissipative pair $(\widehat{\mathcal{P}}(k), \mathcal{Q})$ with the quantum gate coefficient k on the metric graph $\mathbb{Y}$ centered at μ. Notice that $\mathbb{Y}$ can be obtained from the graph $\mathbb{Y}_0$ by a shift.

If the dissipative operator $\widehat{A}$ has no regular points in the upper half-plane, Theorem 12.1 asserts that $\widehat{A}$ is unitarily equivalent the model differentiation operator $\widehat{D} = \widehat{D}_I(0)$ on the graph $\mathbb{Y}_0 = (-\infty, 0) \sqcup (0, \infty)$ in Case (i) with quantum gate coefficient $k = 0$.

The same reasoning as above shows that in this exceptional case, the pair $(\widehat{A}, B)$ is mutually unitarily equivalent to the pair $(\widehat{D}_I(0), \mathcal{Q} - \mu_- \mathcal{R}_- - \mu_+ \mathcal{R}_+)$ on the graph $\mathbb{Y}_0 = (-\infty, 0) \sqcup (0, \infty)$ for some $\mu_\pm \in \mathbb{R}$. Here $\mathcal{R}_-$

and $\mathcal{R}_+$ are the orthogonal projections in

$$L^2(\Upsilon_0) = L^2((-\infty, 0)) \oplus L^2((0, \infty))$$

onto the subspace $L^2((-\infty, 0))$ and $L^2((0, \infty))$, respectively.

If $\mu_+ = \mu_- = \mu$, the pair $(\widehat{D}_I(0), \mathcal{Q} - \mu_-\mathcal{R}_- - \mu_+\mathcal{R}_+) = (\widehat{D}_I(0), \mathcal{Q} - \mu I)$ is mutually unitarily equivalent the canonical dissipative pair $(\widehat{\mathcal{P}}(0), \mathcal{Q})$ on the graph $\Upsilon = (-\infty, \mu) \sqcup (\mu, \infty)$ in Case I with quantum gate coefficient $k = 0$.

If $\mu_+ \neq \mu_-$, then the pair $(\widehat{D}_I(0), \mathcal{Q} - \mu_-\mathcal{R}_- - \mu_+\mathcal{R}_+)$ on the metric graph $\Upsilon_0 = (-\infty, 0) \sqcup (0, \infty)$ (in Case(i)) is mutually unitarily equivalent to the canonical dissipative pair $(\widehat{\mathcal{P}}, \mathcal{Q})$ on the graph $\Upsilon = (-\infty, \mu_-) \sqcup (\mu_+, \infty)$ in Case I*.

The uniqueness part of the statement is an immediate consequence of Lemma 13.11.   □

Remark 13.13. If in addition to the hypotheses of Theorem 13.12 one assumes that A is a self-adjoint (reference) extension of $\dot{A}$, then we immediately get that there exists a self-adjoint extension $\mathcal{P}$ of $\dot{\mathcal{P}}$ such that the quadruple $(\dot{A}, \widehat{A}, A, B)$ is mutually unitarily equivalent to the quadruple $(\dot{\mathcal{P}}, \widehat{\mathcal{P}}(k), \mathcal{P}, \mathcal{Q})$ on the metric graph Υ in Cases I*, I–III with the quantum gate coefficient $k \neq 0$ for some $k \in [0, 1)$. The extension $\mathcal{P}$ is determined by the quadruple $(\dot{A}, \widehat{A}, A, B)$ uniquely unless the graph Υ is in Case I* or in Case I with the quantum gate coefficient $k = 0$.

With a minor modification, the result of Theorem 13.12 extends to the case where the generator $\widehat{A}$ is self-adjoint.

Theorem 13.14. *Assume Hypothesis 13.1. Suppose, in addition, that $\widehat{A} = A$ is self-adjoint and that $\dot{A}$ is a prime symmetric restriction of $\widehat{A}$ with deficiency indices $(1, 1)$ such that*

$$U_t^* \dot{A} U_t = \dot{A} + tI \quad on \ \mathrm{Dom}(\dot{A}). \tag{13.17}$$

Then there exists a unique metric graph $\Upsilon = (-\infty, \mu) \sqcup (\mu, \infty)$ (in Case I) centered at $\mu \in \mathbb{R}$ such that the triple $(\dot{A}, A, B)$ is mutually unitarily equivalent to the Weyl canonical triple $(\dot{\mathcal{P}}, \mathcal{P}, \mathcal{Q})$ on Υ (see Definition 13.9).

Proof. From the hypothesis it follows that in fact (unrestricted) Weyl commutation relations

$$V_s U_t = e^{ist} U_t V_s, \quad t \in \mathbb{R}, \quad s \in \mathbb{R},$$

hold and therefore

$$U_t^* A U_t = A + tI \quad \text{on } \mathrm{Dom}(A), \quad t \in \mathbb{R}. \tag{13.18}$$

By Theorem 3.3, the Weyl-Titchmarsh function associated with the pair $(\dot{A}, A)$ has the form

$$M(z) = i, \quad z \in \mathbb{C}_+.$$

So does the Weyl-Titchmarsh function associated with the pair $(\dot{\mathcal{P}}, \mathcal{P})$ on the graph $\mathbb{Y}_0 = (-\infty, 0) \sqcup (0, \infty)$ in Case (i).

Since $\dot{A}$ and $\dot{\mathcal{P}}$ are prime operators, the pair $(\dot{A}, A)$ is mutually unitarily equivalent to the pair $(\dot{\mathcal{P}}, \mathcal{P})$ on the graph $\mathbb{Y}_0$. That is, there exists a unitary operator $\mathcal{W} : \mathcal{H} \to \mathbb{L}^2(\mathbb{R})$ such that

$$\mathcal{W} A \mathcal{W}^{-1} = \mathcal{P} \quad \text{and} \quad \mathcal{W} \dot{A} \mathcal{W}^{-1} = \dot{\mathcal{P}}.$$

From (13.17) and (13.18) it follows that

$$W_t^* \mathcal{P} W_t = \mathcal{P} + tI \quad \text{on } \mathrm{Dom}(A)$$

and

$$W_t^* \dot{\mathcal{P}} W_t = \dot{\mathcal{P}} + tI \quad \text{on } \mathrm{Dom}(\dot{A}),$$

where

$$W_t = \mathcal{W} U_t^* \mathcal{W}^{-1}.$$

By the definition of the Weyl canonical triple $(\dot{\mathcal{P}}, \mathcal{P}, \mathcal{Q})$, the commutation relations

$$e^{it\mathcal{Q}} \mathcal{P} e^{-it\mathcal{Q}} = \mathcal{P} + tI \quad \text{and} \quad e^{it\mathcal{Q}} \dot{\mathcal{P}} e^{-it\mathcal{Q}} = \dot{\mathcal{P}} + tI$$

hold. Now one can apply Theorem 11.3 to see that there exists a $\mu \in \mathbb{R}$ such that

$$W_t = \mathcal{W} U_t \mathcal{W}^{-1} = e^{-i\mu t} e^{it\mathcal{Q}} = e^{it(\mathcal{Q} - \mu I)}.$$

Therefore, the triple $(\dot{A}, A, B)$ is mutually unitarily equivalent to the triple $(\dot{\mathcal{P}}, \mathcal{P}, \mathcal{Q} - \mu I)$ on the metric graph $\mathbb{Y}_0$.

In turn, the triple $(\dot{\mathcal{P}}, \mathcal{P}, \mathcal{Q} - \mu I)$ is mutually unitarily equivalent to the Weyl canonical triple $(\dot{\mathcal{P}}_\mu, \mathcal{P}, \mathcal{Q})$ on the metric graph $\mathbb{Y}_\mu = (-\infty, \mu) \sqcup (\mu, \infty)$ in Case I. (Recall that $\mathrm{Dom}(\dot{\mathcal{P}}_\mu) = \{ f \in W_2^1(\mathbb{R}) \mid f(\mu) = 0 \}$.)

To complete the proof of the theorem it remains to show that the Weyl triples $(\dot{\mathcal{P}}_\mu, \mathcal{P}, Q)$ and $(\dot{\mathcal{P}}_{\mu'}, \mathcal{P}, Q)$ on the graphs $\mathbb{Y}_\mu$ and $\mathbb{Y}_{\mu'}$, respectively, are not mutually unitarily equivalent unless $\mu = \mu'$. Indeed, assume that they are. Denote by $\mathcal{U}$ the unitary operator that establishes the mentioned mutual unitary equivalence. Since $\mathcal{U}$ commutes with $\mathcal{P}$ and Q, the operator $\mathcal{U}$ is a (unimodular) multiple of the identity. Therefore, $\mathcal{U}^* \dot{\mathcal{P}}_\mu \mathcal{U} = \dot{\mathcal{P}}_{\mu'}$ implies $\dot{\mathcal{P}}_\mu = \dot{\mathcal{P}}_{\mu'}$ and hence $\mu = \mu'$, a contradiction. $\square$

Remark 13.15. Comparing the assumptions of Theorems 13.12 and 13.14 it is clearly seen that the main difference is that in Theorem 13.14 one has to require the commutation relation for the symmetric operator $\dot{A}$, while in the case of Theorem 13.12 the corresponding relations hold automatically. As we have already mentioned in the Introduction, the existence of a symmetric operator $\dot{A}$ with the required properties in the hypothesis of Theorem 13.14 follows from the Stone-von Neumann uniqueness result.

Indeed, let $(\mathcal{P}, \mathcal{Q})$ be the canonical pair with $\mathcal{P} = i\frac{d}{dx}$ and $\mathcal{Q}$ the operator of multiplication by independent variable in $L^2(\mathbb{R})$.

Suppose that $\mathcal{W} : \mathcal{H} \to \mathbb{L}^2(\mathbb{R})$ is a unitary operator such that

$$\mathcal{W}A\mathcal{W}^{-1} = \mathcal{P} \quad \text{and} \quad \mathcal{W}B\mathcal{W}^{-1} = \mathcal{Q}. \tag{13.19}$$

Then the symmetric restriction $\dot{\mathcal{P}}_\mu$ of $\mathcal{P}$ on

$$\text{Dom}(\dot{\mathcal{P}}_\mu) = \{f \in W_2^1(\mathbb{R}) \mid f(\mu) = 0\}$$

has deficiency indices $(1,1)$ and satisfies the commutation relations

$$e^{it\mathcal{Q}}\dot{\mathcal{P}}_\mu e^{-it\mathcal{Q}} = \dot{\mathcal{P}}_\mu + tI.$$

It remains to choose

$$\dot{A} = \mathcal{W}^{-1}\dot{\mathcal{P}}_\mu \mathcal{W} \tag{13.20}$$

and the existence of a restriction with the required properties as follows.

From the uniqueness part of Theorem 13.14 it also follows that if a closed symmetric restriction $\dot{\mathcal{P}}$ of $\mathcal{P}$ (with deficiency indices $(1,1)$) satisfies commutation relations

$$e^{it\mathcal{Q}}\dot{\mathcal{P}}e^{-it\mathcal{Q}} = \dot{\mathcal{P}} + tI,$$

then $\dot{\mathcal{P}} = \dot{\mathcal{P}}_\mu$ for some $\mu \in \mathbb{R}$ (cf. [50]). In this sense as far as the unitary equivalence (13.19) is established (based on the Stone-von Neumann uniqueness result), the choice of $\dot{A}$ via (13.20) in the hypothesis of Theorem 13.14 is canonical.

Notice that as long as the existence of such a restriction is established/required the reasoning above can be considered an independent proof of the Stone-von Neumann uniqueness result.

The following corollary can be considered an *important extension of the Stone-von Neumann uniqueness theorem.*

Corollary 13.16. *Suppose that strongly continuous groups of unitary operators $V_t = e^{iAt}$ and $U_t = e^{iBt}$ in the Hilbert space $\mathcal{H}$ solve the Weyl commutation relations*

$$V_s U_t = e^{ist} U_t V_s, \quad s, t \in \mathbb{R}.$$

Assume that the self-adjoint operator A has simple spectrum. Without loss of generality suppose that $\dot{A}$ is a closed symmetric restriction of A with deficiency indices $(1, 1)$ such that

$$U_t^* \dot{A} U_t = \dot{A} + tI \quad on \ \mathrm{Dom}(\dot{A}), \quad t \in \mathbb{R}. \tag{13.21}$$

If A' is any other self-adjoint extension of $\dot{A}$, then the Weyl commutation relations

$$V_s' U_t = e^{ist} U_t V_s', \quad with \ V_s' = e^{isA'}, \quad s, t \in \mathbb{R}, \tag{13.22}$$

hold.

Proof. By Theorem 13.14, there exists a unique metric graph $\mathbb{Y} = (-\infty, \mu) \sqcup (\mu, \infty)$ such that the triple $(\dot{A}, A, B)$ is mutually unitarily equivalent to the Weyl canonical triple $(\dot{\mathcal{P}}, \mathcal{P}, \mathcal{Q})$ on $\mathbb{Y}$, so that

$$\dot{A} = \mathcal{U} \dot{\mathcal{P}} \mathcal{U}^{-1}, \quad A = \mathcal{U} \mathcal{P} \mathcal{U}^{-1} \quad and \quad B = \mathcal{U} \mathcal{Q} \mathcal{U}^{-1}$$

for some unitary map $\mathcal{U}$ from $L^2(\mathbb{Y})$ onto $\mathcal{H}$. Let A' be a self-adjoint extension of $\dot{A}$. Therefore, $\mathcal{P}' = \mathcal{U}^{-1} A' \mathcal{U}$ is a self-adjoint extension of $\dot{\mathcal{P}}$ on

$$\mathrm{Dom}(\mathcal{P}') = \{ f \in W_2^1((-\infty, \mu)) \oplus W_2^1((\mu, \infty)) \mid f(\mu_-) = \Theta f(\mu_+) \}$$

for some $|\Theta| = 1$.

We have

$$e^{it\mathcal{Q}} \mathcal{P}' e^{-it\mathcal{Q}} = \mathcal{P}' + tI \quad on \ \mathrm{Dom}(\mathcal{P}'), \quad t \in \mathbb{R}, \tag{13.23}$$

and therefore

$$U_t^* A' U_t = \dot{A}' + tI \quad on \ \mathrm{Dom}(A'), \quad t \in \mathbb{R}, \tag{13.24}$$

which in turn implies (13.22). $\qquad\square$

Remark 13.17. In the situation in question one can state more, cf. Remark 13.13. For instance, there exists a unique $\mu \in \mathbb{R}$ and a unique $\Phi \in [0, 2\pi)$ such that the quadruples $(\dot{A}, A, A', B)$ and $(\dot{\mathcal{P}}, \mathcal{P}, \mathcal{P}', \mathcal{Q})$ are mutually unitarily equivalent.

Here $\dot{\mathcal{P}}, \mathcal{P}$ and $\mathcal{P}'$ are differentiation operators in $L^2(\mathbb{R})$ defined on

$$\mathrm{Dom}(\dot{\mathcal{P}}) = \{f \in W_2^1(\mathbb{R}) \,|\, f(\mu) = 0\},$$

$$\mathrm{Dom}(\mathcal{P}) = W_2^1(\mathbb{R}),$$

$$\mathrm{Dom}(\mathcal{P}') = \{f \in W_2^1((-\infty, 0)) \oplus W_2^1((0, \infty)) \,|\, f(\mu+) = e^{i\Phi} f(\mu-)\},$$

respectively, and $\mathcal{Q}$ is the operator of multiplication by independent variable in $L^2(\mathbb{R})$.

More generally, if $\widehat{A}$ is a maximal dissipative extension of $\dot{A}$, then the restricted Weyl commutation relations

$$\widehat{V}_s U_t = e^{ist} U_t \widehat{V}_s, \quad \text{with } \widehat{V}_s = e^{is\widehat{A}}, \quad t \in \mathbb{R}, \quad s \geq 0,$$

hold.

In this case, there exists a unique point $(\mu, \Phi, k) \in \mathbb{R} \times [0, 2\pi) \times [0, 1)$ such that the quadruples $(\dot{A}, \widehat{A}, A', B)$ and $(\dot{\mathcal{P}}, \widehat{\mathcal{P}}, \mathcal{P}', \mathcal{Q})$ are mutually unitarily equivalent. Here the operators $\dot{\mathcal{P}}, \mathcal{P}'$ and $\mathcal{Q}$ are as above and $\widehat{\mathcal{P}}$ is differentiation operators in $L^2(\mathbb{R})$ on

$$\mathrm{Dom}(\widehat{\mathcal{P}}) = \{f \in W_2^1((-\infty, \mu)) \oplus W_2^1((\mu, \infty)) \,|\, f(\mu+) = k f(\mu-)\}.$$

To be complete, we provide the description of the dynamics associated with the strongly continuous semi-group $V_s = e^{i\widehat{\mathcal{P}}s}$ generated by the dissipative momentum operator $\widehat{\mathcal{P}} = \widehat{\mathcal{P}}(k)$ with the quantum gate coefficient $k \in [0, 1)$ on a metric graph $\mathbb{Y}$ in Cases I*, I–III (see Chapter 8 for a more informal description of the dynamics).

Theorem 13.18. *Let $\widehat{\mathcal{P}} = \widehat{\mathcal{P}}(k)$ be the canonical dissipative momentum operator with the quantum gate coefficient $0 \leq k < 1$ on a metric graph $\mathbb{Y}$ in one of the Cases I*, I–III. Then the strongly continuous semi-group $\widehat{V}_s$ of contractions generated by $\widehat{\mathcal{P}}(k)$ in the Hilbert space $L^2(\mathbb{Y})$ admits the following explicit description.*

In Case I*, the semigroup $\widehat{V}_s$ acts as the right shift on the semi-axis $[\mu, \infty)$ and as the truncated right shift on $(-\infty, \nu]$

$$(\widehat{V}_s F)_+(x) = \chi_{[s+\mu, \infty)}(x) f_+(x - s), \quad x \in [\mu, \infty),$$

and

$$(\widehat{V}_s F)_-(x) = f_-(x - s), \quad x \in (-\infty, \nu],$$

where

$$F = (f_-, f_+)^T \in L^2(\mathbb{Y}) = L^2((-\infty, \nu)) \oplus \mathbb{L}^2((\mu, \infty)).$$

In Case I, we have that

$$(\widehat{V}_s F)(x) = (\chi_{(-\infty, \mu)}(x) + k\chi_{[\mu, \infty)}(x)) f_\infty(x - s), \quad x \in \mathbb{R},$$
$$F = f_\infty \in L^2(\mathbb{Y}) = L^2((-\infty, \mu)) \oplus \mathbb{L}^2((\mu, \infty)).$$

In Case II, the semi-group V_s is a nilpotent shift with index $\ell = \nu - \mu > 0$ ($V_s = 0$ for $s \geq \ell$)

$$(\widehat{V}_s F)_\ell(x) = \chi_{[\mu+s, \nu]}(x) f_\ell(x), \quad x \in [\mu, \nu],$$
$$F = f_\ell \in L^2(\mathbb{Y}) = L^2((\mu, \nu)).$$

In Case III, the action of the semi-group $\widehat{V}_s$ is given by

$$(\widehat{V}_s F)_\infty(x) = (\chi_{(-\infty, \mu)}(x) + k\chi_{[\mu, \infty)}(x)) f_\infty(x - s), \quad x \in \mathbb{R},$$
$$(\widehat{V}_s F)_\ell(x) = \chi_{[\mu+s, \nu]}(x) f_\ell(x - s) + \sqrt{1 - k^2} f_\infty(x - s), \quad x \in [\mu, \nu],$$

where

$$F = (f_\infty, f_\ell)^T \in L^2(\mathbb{Y}) = L^2((-\infty, \mu)) \oplus L^2((\mu, \infty)) \oplus L^2((\mu, \nu))$$

and $\ell = \nu - \mu$.

Proof. If the metric graph $\mathbb{Y}$ is in Cases I* or II, there is nothing to prove.
 We provide a complete proof when $\mathbb{Y}$ is in Case I.
 Without loss we may assume that $\mu = 0$. From the definition of the semi-group V_s it follows that

$$\lim_{s \downarrow 0} s^{-1}(\widehat{V}_s - I) f = -f' = i\widehat{\mathcal{P}} f, \quad f \in C_0^\infty(\mathbb{R} \setminus \{0\}). \tag{13.25}$$

Introduce the functions

$$g(x) = \begin{cases} ke^{-x}, & x \geq 0 \\ e^{x}, & x < 0 \end{cases} \quad \text{and} \quad h(x) = \begin{cases} ke^{-x}, & x \geq 0 \\ -e^{x}, & x < 0 \end{cases}.$$

Clearly, $g \in \text{Dom}(i\widehat{\mathcal{P}})$ and $i\widehat{\mathcal{P}}g = h$. It is sufficient to show that

$$\lim_{s \downarrow 0} s^{-1}(\widehat{V}_s - I)g = h = i\widehat{\mathcal{P}}g. \qquad (13.26)$$

Indeed, (13.25) and (13.26) mean that the generator of V_s restricted on the dense linear set $\mathcal{D} = C_0^\infty(\mathbb{R}\backslash\{0\}) + \text{span}\{g\}$ coincides with the operator $\widehat{\mathcal{P}}|_{\mathcal{D}}$. Since the generator is a closed operator and the closure of $\widehat{\mathcal{P}}|_{\mathcal{D}}$ coincides with $\widehat{\mathcal{P}}$ one proves that $\widehat{\mathcal{P}}$ is the generator of the semi-group $\widehat{V}_s$.

Thus, it remains to prove (13.26).

We proceed as follows. Note that

$$(\widehat{V}_s g)(x) = \begin{cases} e^{x-s}, & x < 0 \\ ke^{x-s}, & 0 \leq x < s, \\ e^{s-x}, & x - s \geq 0 \end{cases} \quad \text{a. e. } x \in \mathbb{R},$$

and hence

$$
\begin{aligned}
\|s^{-1}(\widehat{V}_s g - g) - h\|^2 &= \int_{-\infty}^{0} |s^{-1}(e^{x-s} - e^x) + e^x|^2 dx \\
&\quad + \int_{0}^{s} |s^{-1}(ke^{x-s} - ke^x) - ke^{-x}|^2 dx \\
&\quad + \int_{s}^{\infty} |s^{-1}(ke^{s-x} - ke^{-x}) - ke^{-x}|^2 dx \\
&= \left(\frac{e^{-s} - 1 + s}{s}\right)^2 \int_{-\infty}^{0} e^{2x} dx + s^{-2}k^2 \\
&\quad \times \int_{0}^{s} e^{2x} |(e^{-s} - 1 - se^{-2x}|^2 dx \\
&\quad + k\left(\frac{e^s - 1 - s}{s}\right)^2 \int_{s}^{\infty} e^{-2x} dx \to 0 \quad \text{as } s \to 0,
\end{aligned}
$$

proving (13.26).

The proof is complete.　　　　　　　　　　　　　　　　　　　□

Remark 13.19. Taking into account that the generator of the group V_s in Case III is an operator coupling of the ones in Cases I and II, cf. Theorem 8.3, the restriction of the dissipative dynamics in case III to its invariant subspace $L^2((\mu, \nu))$ (with $\nu = \mu + \ell > \mu$) gives rise to the dissipative dynamics in Case II, while the compression $P_{\mathcal{H}} V_s|_{\mathcal{H}}$ of the dissipative dynamics V_s onto its coinvariant subspace $\mathcal{H} = L^2((-\infty, \mu)) \oplus L^2((\mu, \infty))$ leads to the dissipative dynamics in Case I.

One can also compress the dynamics to the channel $L^2((-\infty, \nu)) \oplus L^2((\mu, \infty))$ to obtain the semi-group $\widehat{V}_s$ in Case I*, provided that $\nu < \mu$.

Chapter 14

UNITARY DYNAMICS ON THE FULL GRAPH

The metric graph $\mathbb{Y}$ given by (13.2) can naturally be considered a subgraph of the full metric graph $\mathbb{X} = \mathbb{Y}_\mu \sqcup \mathbb{Y}_\mu$ composed of two identical copies of the metric graph $\mathbb{Y}_\mu = (-\infty, \mu) \sqcup (\mu, \infty)$.

In turn, the dynamics on the metric graph $\mathbb{Y}$ can be dilated to the unitary groups $\widehat{V}_s$ and $\widehat{U}_t$ generated by the canonical variables on $\mathbb{X}$.

To be more precise, we proceed as follows.

In the Hilbert space $L^2(\mathbb{X})$ introduce the unitary group of shifts V_t that can be recognized as evolution operator that maps the initial values of the solution of the following first order hyperbolic system

$$\partial_t \begin{pmatrix} u \\ v \end{pmatrix} + \partial_x \begin{pmatrix} u \\ v \end{pmatrix} = 0 \tag{14.1}$$

with the boundary condition at the vertex $x = \mu$

$$\begin{pmatrix} u \\ v \end{pmatrix}(\mu+) = \begin{pmatrix} k & -\sqrt{1-k^2} \\ \sqrt{1-k^2} & k \end{pmatrix} \begin{pmatrix} u \\ v \end{pmatrix}(\mu-) \tag{14.2}$$

into their value at t. Here k is a parameter such that $0 \le k \le 1$.

The action of the group V_t can easily be described explicitly.

If $t > 0$, one gets that

$$\left(V_t \begin{pmatrix} u \\ v \end{pmatrix} \right)(x) = \begin{pmatrix} u(x-t) \\ v(x-t) \end{pmatrix} \tag{14.3}$$

whenever $x < \mu$ or $t \leq x + \mu$, and

$$\left(V_t \begin{pmatrix} u \\ v \end{pmatrix}\right)(x) = \begin{pmatrix} k\,u(x-t) - \sqrt{1-k^2}\,v(x-t) \\ \sqrt{1-k^2}\,u(x-t) + k\,u(x-t) \end{pmatrix} \qquad (14.4)$$

whenever $\mu \leq x < t$.

If $t < 0$, one clearly has that

$$V_t = (V_{-t})^*, \quad t < 0.$$

The self-adjoint generator $\mathcal{P}$ of the group is the following self-adjoint realization of the differentiation operator on the full graph $\mathbb{X}$ on $\mathrm{Dom}(\mathcal{P})$ consisting of the two-component functions $f = (f_\uparrow, f_\downarrow)^T$,

$$(f_\uparrow, f_\downarrow) \in \left[W_2^1((-\infty, \mu)) \oplus W_2^1((\mu, \infty))\right] \oplus \left[W_2^1((-\infty, \mu)) \oplus W_2^1((\mu, \infty))\right]$$

that satisfy the boundary conditions

$$\begin{pmatrix} f_\uparrow(\mu+) \\ f_\downarrow(\mu+) \end{pmatrix} = \begin{pmatrix} k & -\sqrt{1-k^2} \\ \sqrt{1-k^2} & k \end{pmatrix} \begin{pmatrix} f_\uparrow(\mu-) \\ f_\downarrow(\mu-) \end{pmatrix}. \qquad (14.5)$$

We will call the generator $\mathcal{P}$ the *self-adjoint momentum operator with the quantum gate coefficient* $0 \leq k \leq 1$ on the full graph $\mathbb{X}$.

The unitary dynamics V_t can be illustrated on the example of wave propagation along the transmission line (see (14.2) and also Fig. 14.1 below) and can be described informally as follows.

The wave packet initially located on $(-\infty, \mu)$ in the upper channel is transmitted to the upper channel of the semi-axis (μ, ∞) with the quantum gate coefficient k. The "rest" of the packet gets amplified by $\sqrt{1-k^2}$ and then is transmitted to the lower channel of the semi-axis (μ, ∞). The wave packet on $(-\infty, \mu)$ in the lower channel gets amplified by the factor

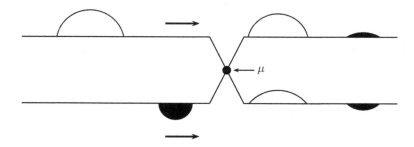

Fig. 14.1. Unitary Dynamics on the Full Metric Graph $\mathbb{X}$

$-\sqrt{1-k^2}$ and by the quantum gate coefficient k and then is transmitted to the upper and lower channel of the semi-axis (μ, ∞), respectively.

Based on the explicit description of the semi-group of contractions $\widehat{V}_s$ provided by Lemma 13.18, we arrive to the main result of this chapter that shows that the dissipative dynamics $\widehat{V}_s$, $s \geq 0$, on the metric graph $\mathbb{Y}$ in any of the Cases I*, I–III can be dilated to a unitary one in the Hilbert space $L^2(\mathbb{X})$ where $\mathbb{X}$ is the full graph. Equivalently, any solution to the restricted Weyl commutation relations

$$U_t \widehat{V}_s = e^{ist} \widehat{V}_s U_t, \quad s \geq 0, \ t \in \mathbb{R},$$

such that the generator of the semi-group $\widehat{V}_s$ belongs to the class $\mathfrak{D}(\mathcal{H})$ (see Appendix G for the definition of the class $\mathfrak{D}(\mathcal{H})$) is unitarily equivalent to a compression of the canonical solution to the Weyl commutation relations

$$U_t V_s = e^{ist} V_s U_t, \quad s, t \in \mathbb{R},$$

in $L^2(\mathbb{R}, \mathbb{C}^2)$ onto an appropriate coinvariant subspace $K \subset L^2(\mathbb{R}, \mathbb{C}^2)$ that reduces the multiplication group U_t.

The precise statement is as follows (cf. [55, Theorem 15]).

Theorem 14.1. *Let $\mathbb{Y} \subset \mathbb{X}$ be a metric graph in one of the Cases I*, I–III given by (13.2). Suppose that $(\widehat{\mathcal{P}}(k), \mathcal{Q}(\mathbb{Y}))$ is the canonical dissipative pair with the quantum gate coefficient k on the metric graph $\mathbb{Y}$ and $(\mathcal{P}(k), \mathcal{Q}(\mathbb{X}))$ is the canonical pair on the full graph $\mathbb{X}$.*
Then

$$e^{i\widehat{\mathcal{P}}(k)s} = P_{L^2(\mathbb{Y})} e^{i\mathcal{P}(k)s}|_{L^2(\mathbb{Y})}, \quad s \geq 0, \tag{14.6}$$

and

$$e^{i\mathcal{Q}(\mathbb{Y})t} = P_{L^2(\mathbb{Y})} e^{i\mathcal{Q}(\mathbb{X})t}|_{L^2(\mathbb{Y})}, \quad t \in \mathbb{R}. \tag{14.7}$$

Here $P_{L^2(\mathbb{Y})}$ stands for the orthogonal projection from the space $L^2(\mathbb{X})$ onto the subspace $L^2(\mathbb{Y}) \subset L^2(\mathbb{X})$.

Proof. It is convenient to identify the Hilbert space $L^2(\mathbb{X})$ as the von Neumann integral

$$L^2(\mathbb{X}) = \begin{pmatrix} L^2((-\infty, \mu)) & L^2((\mu, \infty)) \\ L^2((-\infty, \mu)) & L^2((\mu, \infty)) \end{pmatrix}.$$

Since any metric graph $\mathbb{Y}$ in one of the Cases I*, I–III can naturally be considered a subgraph of the full graph $\mathbb{X}$, the Hilbert space $L^2(\mathbb{Y})$ can be identified with

$$L^2(\mathbb{Y}) \approx \begin{cases} \begin{cases} \begin{pmatrix} L^2((-\infty,\nu)), & L^2((\mu,\infty)) \\ 0 & 0 \end{pmatrix}, & (\nu < \mu) \\ \begin{pmatrix} L^2((-\infty,\mu)) & L^2((\mu,\infty)) \\ 0 & L^2((\mu,\nu)) \end{pmatrix}, & (\nu \geq \mu) \end{cases} & \text{in Case I*} \\[1em] \begin{pmatrix} L^2((-\infty,\mu)) & L^2((\mu,\infty)) \\ 0 & 0 \end{pmatrix}, & \text{in Case I} \\[1em] \begin{pmatrix} 0 & 0 \\ 0 & L^2((\mu,\nu)) \end{pmatrix}, & \text{in Case II} \\[1em] \begin{pmatrix} L^2((-\infty,\mu)) & L^2((\mu,\infty)) \\ 0 & L^2((\mu,\nu)) \end{pmatrix}, & \text{in Case III.} \end{cases}$$

Clearly, the subspace $L^2(\mathbb{X}) \ominus L^2(\mathbb{Y})$ splits into the direct sum of incoming $\mathcal{D}_-$ and outgoing subspaces $\mathcal{D}_+$ for the group $\widehat{V}_t$,

$$L^2(\mathbb{X}) \ominus L^2(\mathbb{Y}) = \mathcal{D}_- \oplus \mathcal{D}_+.$$

For instance, in Case III,

$$\mathcal{D}_- = \begin{pmatrix} 0 & 0 \\ L^2((-\infty,\mu)) & 0 \end{pmatrix} \quad \text{and} \quad \mathcal{D}_+ = \begin{pmatrix} 0 & 0 \\ 0 & L^2((\nu,\infty)) \end{pmatrix}.$$

Therefore the restriction of the unitary group $V_t = e^{i\mathcal{P}(k)t}$ onto its coinvariant subspaces $K = L^2(\mathbb{Y})$ is a strongly continuous semi-group of contractions.

Comparing (14.3), (14.4) with the explicit description for the semi-groups $\widehat{V}_s$ action provided by Lemma 13.18 in each of the cases, one gets that the unitary evolution V_s, $s \in \mathbb{R}$, in $L^2(\mathbb{X})$ compressed to the (coinvariant) subspace $L^2(\mathbb{Y})$ gives rise to the corresponding semi-groups of contractions

$$\widehat{V}_s = P_{L^2(\mathbb{Y})} V_s|_{L^2(\mathbb{Y})}, \quad s \geq 0,$$

which proves (14.6). To obtain (14.7) it remains to observe that the subspace $L^2(\mathbb{Y})$ reduces the multiplication group $U_t = e^{-i\mathcal{Q}(\mathbb{X})t}$. □

Part 2

Continuous Monitoring, Quantum Measurements

Chapter 15

CONTINUOUS MONITORING
OF THE QUANTUM SYSTEMS

The aim of this chapter is to present a general discussion of possible outcomes in the frequent quantum measurement theory.

Let $\mathcal{H}$ be the Hilbert space used in the description of a quantum system with the Hamiltonian H, a self-adjoint operator in $\mathcal{H}$. Recall that the time evolution of an initial state ϕ of the system, a unit vector in the Hilbert space $\mathcal{H}$, is described by a one-parameter group of unitary operators $U_t = e^{-it/\hbar H}$. Notice that if $\phi \in \text{Dom}(H)$, the vector $\psi(t) = e^{-it/\hbar H}\phi$ satisfies the time-dependent Schrödinger equation

$$i\hbar \frac{\partial \psi}{\partial t} = H\psi.$$

Let $p(t)$ denote the survival probability

$$p(t) = |(e^{-it/\hbar H}\phi, \phi)|^2 = |(e^{it/\hbar H}\phi, \phi)|^2. \tag{15.1}$$

One of the central problems of frequent quantum measurement theory is to study the time behavior of $[p(t/n)]^m$ for large m and n, and in particular, the computation of the limit

$$p_c(t) = \lim_{n\to\infty} [p(t/n)]^n.$$

We will call $p_c(t)$ the *survival probability under continuous monitoring* of the system and focus on the following three possible scenarios:

α) $p_c(t) = 1$, the quantum Zeno effect;
β) $p_c(t) = 0$, the quantum Anti-Zeno effect;
γ) $p_c(t) = e^{-\tau|t|}$ for some $\tau > 0$, the Exponential Decay.

15.1. Quantum Zeno Effect

Hypothesis 15.1. Suppose that H is a self-adjoint operator in the Hilbert space $\mathcal{H}$ expressed by

$$H = \int_{\mathbb{R}} \lambda dE_H(\lambda),$$

where $E_H(\lambda)$ is a resolution of the identity. Let ϕ be a unit vector (state) in $\mathcal{H}$ and $\nu_\phi(d\lambda)$ denote the spectral measure of the state ϕ,

$$\nu_\phi(d\lambda) = (E_H(d\lambda)\phi, \phi).$$

Assume that $N(\lambda)$ is the corresponding right-continuous distribution function

$$N(\lambda) = \nu_\phi((-\infty, \lambda]), \quad \lambda \in \mathbb{R}. \tag{15.2}$$

Definition 15.2. We say that ϕ is a Zeno state under continuous monitoring of the quantum unitary evolution $\phi \to e^{itH}\phi$ if

$$\lim_{n \to \infty} |(e^{it/nH}\phi, \phi)|^{2n} = 1 \quad \text{for all } t \geq 0.$$

Recall the following necessary and sufficient conditions for the Quantum Zeno effect to occur.

Proposition 15.3 ([6]). *Assume Hypothesis* 15.1. *Then the state ϕ is a Zeno state if and only if the light tails requirement*

$$\lim_{\lambda \to \infty} \lambda(1 - N(\lambda) + N(-\lambda)) = 0 \tag{15.3}$$

holds.

In particular, if $\phi \in \mathrm{Dom}(|H|^{1/2})$, then ϕ is a Zeno state.

Remark 15.4. The discovery of the phenomenon that the evolution of quantum system can be eventually frozen under continuous monitoring is due to A. Turing (Turing's paradox)[1] and L. Khalfin [58] while the term the quantum Zeno effect was coined by B. Misra and E. C. G. Sudarshan [93].

---

[1] In his 1954 letter to M. H. A. Newman, Turing wrote: "it is easy to show using standard theory that if a system starts in an eigenstate of some observable, and measurements are made of that observable N times a second, then, even if the state is not a stationary one, the probability that the system will be in the same state after, say, 1 second tends to one as N tends to infinity; i.e. that continual observation will prevent motion...", see [132].

The necessary and sufficient condition for the occurrence of the quantum Zeno effect is due to H. Atmanspacher, W. Ehm and T. Gneiting [6] where the authors explored the well known fact (see [33, Theorem 1, p. 232]) that the light tails requirement (15.3) is necessary and sufficient for the weak law of large numbers to hold. We also refer to [119] for an excellent introduction to the subject. The quantum Zeno dynamics of a relativistic system is discussed in [91]. As for an experimental confirmation of the effect see [120].

Remark 15.5. Notice that the membership $\phi \in \text{Dom}(|H|^{1/2})$ means that the spectral (probability) measure $(E_H(d\lambda)\phi, \phi)$ has the first moment and hence ϕ is a Zeno state. This can also be seen directly (cf. [58]) as follows.

Consider a sequence $\xi_1, \xi_2, \ldots$ of independent copies of a random variable ξ with the common distribution function $N(\lambda)$ given by (15.2). By the strong law of large numbers,

$$\lim_{n \to \infty} \frac{\xi_1 + \xi_2 + \cdots + \xi_n}{n} = a \quad \text{almost surely,}$$

where

$$a = \mathsf{E}\xi = (\text{sgn}(H)|H|^{1/2}\phi, |H|^{1/2}\phi) \in \mathbb{R}$$

is the mathematical expectation of the random variable ξ.

Recall that $\phi \in \text{Dom}(|H|^{1/2})$ and therefore the right hand side $(\text{sgn}(H)|H|^{1/2}\phi, |H|^{1/2}\phi)$ is well defined. Since the random variables $\xi_1, \xi_2, \ldots$ are independent and equidistributed, we have

$$(e^{it/nH}\phi, \phi)^n = (\mathsf{E}e^{it/n\xi})^n = \underbrace{\mathsf{E}e^{it/n\xi_1} \cdot \mathsf{E}e^{it/n\xi_2} \cdot \ldots \cdot \mathsf{E}e^{it/n\xi_n}}_{n \text{ times}}$$

$$= \mathsf{E}e^{it\frac{\xi_1 + \xi_2 + \cdots + \xi_n}{n}} \to e^{ita} \quad \text{as } n \to \infty,$$

which shows that

$$\lim_{n \to \infty} |(e^{it/nH}\phi, \phi)|^{2n} = 1. \tag{15.4}$$

That is, ϕ is a Zeno state.

15.2. Anti-Zeno Effect

Frequent observations can also accelerate the decay process and the corresponding phenomenon is known as the quantum anti-Zeno effect.

Definition 15.6. We say that ϕ, $\|\phi\| = 1$, is an anti-Zeno state under continuous monitoring of the quantum unitary evolution $\phi \to e^{itH}\phi$ if

$$\lim_{n\to\infty} |(e^{it/nH}\phi, \phi)|^{2n} = 0 \quad \text{for all } t > 0. \tag{15.5}$$

The following lemma provides a simple sufficient condition for a state to be an anti-Zeno state. We state the corresponding result using the language of the theory of limit distributions of sums of independent random variables (we refer to Appendix H for the terminology and a brief exposition of the theory).

Lemma 15.7 (cf. [30]). *Assume Hypothesis* 15.1. *Suppose that the distribution $N(\lambda)$ of the state ϕ belongs to the domain of attraction of an α-stable law with $0 < \alpha < 1$. Then ϕ is an anti-Zeno state.*

Proof. The characteristic function of the distribution $N(\lambda)$ coincides with $(e^{itH}\phi, \phi)$ and therefore admits the representation (by Remark H.2 in Appendix H)

$$(e^{itH}\phi, \phi) = \exp\left(-\sigma|t|^\alpha \tilde{h}(t)(1 - i\beta\frac{t}{|t|}\omega(t, \alpha))\right),$$

where $\tilde{h}(t)$ is slowly varying as $t \to 0$. In particular,

$$|(e^{it/nH}\phi, \phi)|^{2n} = \exp(-2cn^{1-\alpha}\tilde{h}(t/n)|t|^\alpha(1 + o(1))) \quad \text{as } n \to \infty.$$

Since (see, e.g., [48, Appendix 1])

$$\lim_{n\to\infty} n^{1-\alpha}\tilde{h}(t/n) = +\infty,$$

one concludes that

$$\lim_{n\to\infty} |(e^{it/nH}\phi, \phi)|^{2n} = 0, \quad t \neq 0. \qquad \square$$

Remark 15.8. Necessary and sufficient conditions for the quantum anti-Zeno effect to occur can be found in [6, Theorem 2].

Remark 15.9. It is worth mentioning that the situation is quite different if the distribution function belongs to the domain of attraction of an α-stable law with $1 < \alpha \leq 2$. In this case the probability measure $\nu_\phi(d\lambda)$ has the first moment and therefore the state ϕ is a Zeno state.

15.3. The Exponential Decay

Next we turn to the borderline case of α-stable distributions with $\alpha = 1$ which play an exceptional role in explanation of the exponential decay phenomenon under continuous monitoring.

Definition 15.10. We say that ϕ is a *resonant state* under continuous monitoring of the quantum unitary evolution $\phi \to e^{itH}\phi$ if

$$\lim_{n\to\infty} |(e^{it/nH}\phi, \phi)|^{2n} = e^{-\tau|t|}, \quad \text{for some } \tau > 0 \quad \text{and all } t \geq 0. \quad (15.6)$$

Remark 15.11. Notice that one can also consider exponentially decaying (resonant) states by requiring that the survival probability $p(t) = |(e^{itH}\phi, \phi)|^2$ tends to zero exponentially fast as $|t|$ approaches infinity. In this case, however, the spectrum of the Hamiltonian H has to fill in the whole real axis which excludes from the consideration the quantum systems with semi-bounded Hamiltonians.

It can be easily seen as follows. The requirement that the survival probability $p(t)$ falls off exponentially implies that the survival probability amplitude $(e^{itH}\phi, \phi)$ is the Fourier transform of an absolute continuous measure with the density that is analytic in a strip containing the real axis. In particular, the Radon-Nykodim derivative $\frac{d}{d\lambda}(E_H(\lambda)\phi, \phi)$ of the spectral measure of the element ϕ is positive almost everywhere which shows that $\mathrm{spec}(H) = \mathbb{R}$. If the Hamiltonian H has a gap in its spectrum, then there are no exponentially decaying states whatsoever unless the quantum system is under continuous monitoring. The geometric reason behind is that the unitary group e^{itH} does not have orthogonal incoming and outgoing subspaces as it follows from the Hegerfeldt Theorem [45].

A sufficient condition for the exponential decay (15.6) is provided by the following corollary of the Gnedenko-Kolmorogov limit theorem (see Theorem H.1 in Appendix H).

Theorem 15.12. *Assume Hypothesis 15.1. Suppose, in addition, that*

$$\lim_{\lambda\to\infty} \lambda(1 - N(\lambda)) = \frac{1+\beta}{\pi}\sigma \quad \text{and} \quad \lim_{\lambda\to\infty} \lambda N(-\lambda) = \frac{1-\beta}{\pi}\sigma \quad (15.7)$$

for some $\sigma > 0$ and $\beta \in [-1, 1]$. Then ϕ is a resonant state and

$$\lim_{n\to\infty} |(e^{it/nH}\phi, \phi)|^{2n} = e^{-2\sigma|t|}, \quad t \in \mathbb{R}. \quad (15.8)$$

Proof. By Theorem H.1 and Remark H.2 in Appendix H, the distribution $N(\lambda)$ belongs to the domain of normal attraction of the 1-stable law. In particular, there are constants A_n such that

$$\lim_{n\to\infty} (e^{it/nH}\phi, \phi)^n e^{iA_n} = \exp\left(-\sigma|t|\left(1 + i\beta\frac{2}{\pi}\frac{t}{|t|}\log|t|\right)\right),$$

from which (15.8) follows. □

Remark 15.13. If $\sigma > 0$ and therefore ϕ is a resonant state, the probability measure $\nu_\phi(d\lambda)$ does not have the first moment and hence

$$\phi \notin \mathrm{Dom}(|H|^{1/2}).$$

In this case the "total energy" of the quantum system in the state ϕ is infinite. Introducing the "ultra-violet" cut-off Hamiltonian

$$H_E = \int_{|\lambda|\le E} \lambda dE_H(\lambda),$$

one observes that the "truncated" energy $(|H_E|\phi, \phi)$ of the state ϕ is log-divergent as the truncation parameter E approaches infinity. That is,

$$(|H_E|\phi, \phi) = \frac{2}{\pi}\sigma\log E + o(\log E) \quad \text{as } E \to \infty. \tag{15.9}$$

In particular, the parameter σ determines the rate of convergence of the mean-value cut-off energy in the logarithmic scale as

$$\sigma = \frac{\pi}{2}\lim_{E\to\infty}\frac{(|H_E|\phi, \phi)}{\log E}.$$

Indeed, one gets that

$$(H_E\phi, \phi) = \int_{[-E,E]} \lambda dN(\lambda) = \int_{[-E,0)} \lambda dN(\lambda) + \int_{(0,E]} \lambda dN(\lambda). \tag{15.10}$$

Integrating by parts (see, e.g., [37, Theorem 3.36]) one obtains

$$\int_{(0,E]} \lambda dN(\lambda) = \int_{(0,E]} \lambda d(N(\lambda) - 1)$$

$$= E(N(E) - 1) + \int_0^E (1 - N(\lambda))d\lambda$$

$$= -\frac{1+\beta}{\pi}\sigma + \left(\frac{1+\beta}{\pi}\sigma\log E + o(\log E)\right) \quad \text{as } E \to \infty,$$

where we have used (15.7) on the last step.

Therefore,

$$\int_{(0,E]} \lambda dN(\lambda) = \frac{1+\beta}{\pi}\sigma \log E + o(\log E) \quad \text{as } E \to \infty. \tag{15.11}$$

In a similar way one shows that

$$\int_{[-E,0)} \lambda dN(\lambda) = -\frac{1-\beta}{\pi}\sigma \log E + o(\log E) \quad \text{as } E \to \infty, \tag{15.12}$$

which together with (15.11) implies (15.9).

Combining (15.10), (15.11) and (15.12), one also justifies the logarithmic divergence of the averaged truncated energy of the state ($\beta \neq 0$), that is,

$$(H_E\phi, \phi) = \frac{2\beta}{\pi}\sigma \log E + o(\log E) \quad \text{as } E \to \infty. \tag{15.13}$$

Example 15.14. The following example shows that neither the non-semi-boundedness of the operator H nor the requirement of absolutely continuity of its spectrum is necessary for the existence of resonant states under continuous observation.

Let H be a self-adjoint operator in the Hilbert space with simple discrete spectrum such that

$$\text{spec}(H) = \mathbb{N}$$

and ϕ_n the corresponding eigenfunctions,

$$H\phi_n = n\phi_n, \quad n \in \mathbb{N}, \quad \|\phi_n\| = 1.$$

Let ψ be a state

$$\psi = \sum_{n \in \mathbb{N}} a_n\phi_n, \tag{15.14}$$

where $\{a_n\}_{n=1}^{\infty}$ is a sequence of complex numbers such that

$$\sum_{n \in \mathbb{N}} |a_n|^2 = 1.$$

Then

$$N(\lambda) = \sum_{n < \lambda} |a_n|^2.$$

In the special case

$$a_n = \frac{1}{\sqrt{\zeta(2s)}} \frac{1}{n^s}, \quad \frac{1}{2} < s,$$

where $\zeta(s)$ is the Riemann zeta-function, denote by ψ_s the corresponding state given by (15.14). Then for the distribution function $N_s(\lambda)$ associated with the state ψ_s the following asymptotics

$$1 - N_s(\lambda) = \frac{1}{\zeta(2s)(2s-1)} \frac{1}{\lambda^{2s-1}}(1 + o(1)) \quad \text{as } \lambda \to \infty$$

holds. In particular, since $N_s(\lambda) = 0$ for $\lambda < 1$, the law $N_s(\lambda)$ belongs to the domain of attraction of a stable law with exponent $2s - 1$ for $s \in \left(\frac{1}{2}, \frac{3}{2}\right]$ by Theorem H.1. Moreover,

$$\psi_s \in \text{Dom}(H^{1/2}), \quad s > 1.$$

Combining Lemma 15.7, Theorem 15.12 and Proposition 15.3 (cf. Remark 15.5) one concludes that the state

$$\psi_s = \frac{1}{\sqrt{\zeta(2s)}} \sum_{n \in \mathbb{N}} \frac{1}{n^s}\phi_n \quad \text{is} \quad \begin{cases} \text{an anti-Zeno state,} & \frac{1}{2} < s < 1 \\ \text{a resonant state,} & s = 1 \\ \text{a Zeno state,} & 1 < s \end{cases} \quad (15.15)$$

In the case $s = 1$, the distribution function asymptotics simplifies to

$$1 - N_1(\lambda) = \frac{6}{\pi^2} \frac{1}{\lambda}(1 + o(1)) \quad \text{as } \lambda \to \infty.$$

Then, from Theorem 15.12 it follows that the resonant state ψ_1 exponentially decays under continuous monitoring and

$$\lim_{n \to \infty} |(e^{it/nH}\psi_1, \psi_1)|^{2n} = e^{-\frac{6}{\pi}|t|}, \quad t \in \mathbb{R}. \quad (15.16)$$

Notice that the operator H can be realized in the Hilbert space $L^2(\mathbb{R})$ as the Hamiltonian of the familiar quantum harmonic oscillator

$$H = -\frac{1}{2}\frac{d^2}{dx^2} + \frac{x^2 + 1}{2}$$

with the domain

$$D(H) = W^{2,2}(\mathbb{R}) \cap L^2(\mathbb{R}; (1 + x^2)^2 dx).$$

The corresponding eigenfunctions ϕ_n are known to be given by

$$\phi_{k+1}(x) = \sqrt[4]{\frac{1}{\pi}} \frac{1}{\sqrt{2^k k!}} e^{-\frac{1}{2}x^2} H_k(x), \quad k = 0, 1, 2, \ldots,$$

where $H_k(x)$, are classical Hermite-Tchebyscheff polynomials (see, e.g., [131, Ch. 2 Theorem 2.1]).

Summing up, the quantum oscillator prepared initially in the pure state

$$\psi_1 = \frac{\sqrt{6}}{\pi} \sum_{n \in \mathbb{N}} \frac{1}{n} \phi_n$$

is an example of a quantum system that exhibits exponential decay under continuous monitoring, while it Hamiltonian is a positive operator with a discrete spectrum. Notice that the distribution function $N_1(\lambda)$ belongs to the domain of normal attraction of the 1-stable law, the Landau distribution (see, e.g., [76, Ch. 7]), the characteristic function $g(t)$ of which is given by

$$g(t) = e^{-\frac{3}{\pi}|t|\left(1+i\frac{2}{\pi}\log|t|\right)}, \quad t \in \mathbb{R}.$$

This example of a quantum oscillator (pendulum) brings to mind a charming metaphor: *Under continuous monitoring, Schrödinger's Cat, swinging on the quantum swing, can easily turn into the gradually disappearing Cheshire Cat, leaving us alone to enjoy its vanishing (ironic) grin.*

We conclude this subsection by an abstract result that takes place for a special class of hyperbolic (quantum) systems the Hamiltonian of which is a self-adjoint dilation of a dissipative operator (cf. (5.22) for the definition of a self-adjoint dilation).

Lemma 15.15. *Suppose that a self-adjoint operator H in a Hilbert space $\mathcal{H}$ dilates a maximal dissipative operator $\widehat{A}$ acting in a subspace $\mathcal{K} \subset \mathcal{H}$. Assume that $\phi \in \mathrm{Dom}(\widehat{A}) \subset \mathcal{K}$ is such that $\|\phi\| = 1$.*
Then

$$\lim_{n \to \infty} |(e^{it/nH}\phi, \phi)|^{2n} = \lim_{n \to \infty} |(e^{it/n\widehat{A}}\phi, \phi)|^{2n} = e^{-\tau t}, \quad t \geq 0, \qquad (15.17)$$

where

$$\tau = 2\mathrm{Im}(\widehat{A}\phi, \phi).$$

Proof. Since $\phi \in \mathrm{Dom}(\widehat{A})$, we have

$$(e^{i\varepsilon\widehat{A}}\phi, \phi) = 1 + i\varepsilon(\widehat{A}\phi, \phi) + o(\varepsilon) \quad \text{as } \varepsilon \downarrow 0.$$

Therefore,

$$|(e^{i\varepsilon\widehat{A}}\phi, \phi)| = 1 - \varepsilon\mathrm{Im}(\widehat{A}\phi, \phi) + o(\varepsilon) \quad \text{as } \varepsilon \downarrow 0,$$

and hence

$$\lim_{n \to \infty} |(e^{it/n\widehat{A}}\phi, \phi)|^{2n} = e^{-2\mathrm{Im}(\widehat{A}\phi, \phi)t}, \quad t \geq 0.$$

By the hypothesis, the Hamiltonian H dilates $\widehat{A}$ and hence

$$\lim_{n\to\infty} |(e^{it/nH}\phi,\phi)|^{2n} = \lim_{n\to\infty} |(e^{it/n\widehat{A}}\phi,\phi)|^{2n} = e^{-\tau t}, \quad t \geq 0.$$

$\square$

15.4. Frequent Measurements and the Time-Energy Uncertainty Principle

The study of the limit behavior of the survival probability $[p_c(t/n)]^m$ as $m, n \to \infty$ and $m \neq n$ in appropriate time-scales is also of definite interest. For instance, if $\frac{m}{n} \to \infty$, we deal with the case of *prolonged* frequent quantum measurements.

For instance, prolonged frequent measurements with $m(n) = n^2$ can eventually "unfreeze" states that were *a priory* Zeno states.

Indeed, assume that $\phi \in \mathrm{Dom}(H)$ and therefore ϕ is a Zeno state. We have

$$(e^{itH}\phi,\phi) = 1 + i(H\phi,\phi)t - \frac{1}{2}(H\phi,H\phi)t^2 + o(t^2), \quad \text{as } t \to 0.$$

In particular,

$$|(e^{itH}\phi,\phi)|^2 = 1 - t^2(\Delta H)^2 + o(t^2),$$

where

$$\Delta H = \left(\|H\phi\|^2 - (H\phi,\phi)^2\right)^{1/2} < \infty.$$

Therefore,

$$\lim_{n\to\infty} |(e^{it/nH}\phi,\phi)|^{2n^2} = e^{-(\Delta H)^2 t^2} \tag{15.18}$$

and the state ϕ exhibits an exponential decay in a non-linear time scale. In other words, if the quantum system is observed n^2 times with the "frequency" $\omega = n/t$ (t is fixed and n is large) the results of the *prolonged frequent measurements* of the survival probability unfreezes the Zeno state $\phi \in \mathrm{Dom}(H)$.

More precisely, the state ϕ is a resonant state in the time-scale

$$\mathfrak{t}(t) = \sqrt{t}$$

in the sense that

$$\lim_{n\to\infty} [p(\mathfrak{t}(t/n))]^n = \exp\left(-(\Delta H)^2|t|\right). \tag{15.19}$$

In this situation one can give an estimate for the survival probability from above via the angle $\theta(t) = \arccos|(e^{itH}\phi, \phi)|$ between the states ϕ and $e^{itH}\phi$, which is an important geometric characteristics of the trajectory $\psi(t) = e^{itH}\phi$ in the Hilbert space.

To do so, we use the Mandelstam-Tamm time-energy uncertainty relation [90] (also see [110])

$$\theta(t) \leq t\Delta H \quad \left(0 \leq t\Delta H \leq \frac{\pi}{2}\right), \tag{15.20}$$

one gets an *a posteriori* estimate

$$\lim_{n\to\infty} |(e^{it/nH}\phi, \phi)|^{2n^2} \leq e^{-\theta^2(t)}$$

via the angle $\theta(t)$.

15.5. Presto

In a more general setting, see, e.g., [29, 93], assume that $\mathfrak{S}$ is a quantum system and $\mathcal{H}$ the Hilbert space of (pure) states of the system $\mathfrak{S}$. Assume that the system is in the state given by a density operator ρ, with ρ a non-negative trace class operator with trace one. Suppose that the dynamics of $\mathfrak{S}$ is goverened by the strongly continuous group of unitary operators $U(t)$ such that the time dependence of states is given by

$$\rho \longrightarrow \rho_t = U(t)\rho U(t)^* \quad \text{(Schrödinger's picture)}.$$

Let P be an orthogonal projector onto the subspace spanned by the unde-cayed (unstable) states, the observable, which can also be understood a quantum statement or event corresponding to the "yes-no-experiment" [49, Ch. 5]. The probability of the event P to occur at the instant t is then given by

$$p(t) = \text{tr}(\rho_t P) = \text{tr}(U(t)\rho U(t)^* P). \tag{15.21}$$

We remark that if ρ is a pure state

$$\rho = (\cdot, \phi)\phi, \quad \|\phi\| = 1,$$

and P is a one-dimensional projection corresponding to the unit vector ϕ, then $p(t)$ given by (15.21) coincides with the survival probability (15.1).

Notice that being observed $(n + 1)$ times at the moments of time $0, t/n, 2t/n, \ldots, t$, the system finds itself in one of the states from $\text{Ran}(P)$

with probability

$$p(t, n) = \text{tr}(T_n(t)\rho T_n^*(t)),$$

where

$$T_n(t) = [PU(t/n)P]^n.$$

Under the hypothesis that the limit

$$T(t) = \operatorname*{s-lim}_{n \to \infty} T_n(t), \quad t \geq 0, \tag{15.22}$$

exists and forms a strongly continuous semigroup of contractions, denote by B the generator of the semigroup $T(t)$, that is,

$$T(t) = e^{itB}, \quad t \geq 0. \tag{15.23}$$

Denote by P_s the orthogonal projection in the Hilbert space $\text{Ran}(P)$ such that the part of B in $\text{Ran}(P_s)$ is a self-adjoint operator and the restriction of B onto $\text{Ran}(P_d)$ with $P_d = P \ominus P_s$, is a completely non-selfadjoint dissipative operator. The study of the open system $\mathfrak{S}'$ in the state

$$\rho' = \frac{P_d \rho P_d}{\text{tr}(\rho P_d)}$$

governed by the evolution semigroup $T(t)$,

$$\rho' \longrightarrow \rho'_t = \widehat{T}(t)\rho'\widehat{T}(t)^*,$$

where

$$\widehat{T}(t) = P_d T(t)|_{\text{Ran}(P_d)}$$

provides an interesting example of a dissipative system which serves as a model for quantum systems under continuous observation.

Despite the obvious mathematical attractiveness, the hypotheses (15.22) and (15.23) are incompatible with the physical requirement that the Hamiltonian of the system has to be semibounded below.

To see that suppose that ρ is a pure state,

$$\rho = (\cdot, \phi)\phi,$$

and that $P = \rho$. In this case the hypothesis that $P_d \neq 0$ boils down to the existence of the limit

$$\lim_{n \to \infty} [(U(t/n)\phi, \phi)]^n = e^{i\alpha t}, \quad t \geq 0 \quad (\text{Im}(\alpha) > 0). \tag{15.24}$$

In particular, the distribution function $N(\lambda)$ of the state ϕ given by (15.2) belongs to the domain of the normal attraction of the 1-stable law, the density of which is given by the Breit-Wigner-Cauchy-Lorentz shape

$$\varrho(d\lambda) = \frac{\sigma}{2\pi} \frac{1}{\lambda^2 + \frac{1}{4}\sigma^2} d\lambda, \quad \text{with } \sigma = \text{Im}(\alpha).$$

By Theorem 15.12,

$$\lim_{\lambda \to +\infty} \lambda(1 - N(\lambda)) = \lim_{\lambda \to +\infty} \lambda N(-\lambda) = \frac{\sigma}{\pi} > 0,$$

which shows that the Hamiltonian H of the system is neither semibounded from below nor from above.

As a corollary, one immediately concludes that for the quantum oscillator in Example 15.14 in the state ϕ given by (15.15) for $s = 1$, the limit

$$T(t) = \lim_{n \to \infty} (U(t/n)\phi, \phi)^n$$

does not exists although the limit

$$p(t) = \lim_{n \to \infty} |(U(t/n)\phi, \phi)|^{2n} = \lim_{n \to \infty} |(e^{it/nH}\phi, \phi)|^{2n}$$

is well defined. Notice that in Example 15.14 the Hamiltonian H of the quantum oscillator is a positive operator with discrete spectrum and ϕ is a resonant state (see Definition 15.10) the energy distribution of which belongs to the domain of attraction of the Landau distribution, not to the Cauchy one.

Remark 15.16. We remark that if (15.24) holds, then the generator of the corresponding contraction semigroup $\widehat{T}(t)$ can be identified with the complex number α from the open upper half-plane. In particular, the (classical) dissipative system

$$\frac{1}{i}\frac{d}{dt}y = \alpha y, \quad y(0) = 1,$$

can be dilated ("quantized") to the closed quantum system in $L^2(\mathbb{R})$ initially prepared in the pure state

$$\phi(x) = \frac{1}{\sqrt{2\text{Im}(\alpha)}} \begin{cases} 0, & x > 0 \\ e^{-i\alpha x}, & x \leq 0 \end{cases}$$

time evolution of which is given by the right shift operator $V(t)$ in $L^2(\mathbb{R})$,

$$(V(t)\phi)(x) = \phi(x - t).$$

It is interesting to notice that the generator $i\frac{d}{dx}$ of the shift group $V(t)$ together with the multiplications operator by independent variable x solve the canonical commutation relations with which we began to present the results of our monograph.

We conclude this chapter with the following informal discussions.

Perhaps the easiest informal way to understand the presence of exponentially decaying terms in the dynamics of open quantum system is as follows.

Under certain conditions, the compressed resolvent of the Hamiltonian (onto the "unstable" subspace $\text{Ran}(P)$) admits an analytic continuation to the "unphysical sheet" where it may have poles type singularities (resonances). Representing the reduced evolution operator of the open quantum system as the Riesz type contour integral via the compressed resolvent $P(H - \lambda I)^{-1}P$, one can deform the enclosing spectrum integration contour to the non-physical sheet. Evaluating the integral by the residue theorem yields exponentially decreasing terms in the description of the unitary reduced dynamics. In this schema, which is essentially due to Gamow, the analytic continuation ansatz necessarily requires the spectrum of the Hamiltonian to possess an absolutely continuous component. However, if the Hamiltonian of the system is a semi-bounded operator, the contour integration around the threshold of the continuous spectrum, the branching point of the resolvent kernel, gives rise to power law decay, which one again confirms that that there is no exponentially decay in the quantum systems with a semibounded Hamiltonian if one decides to start from the "first principles".

Under more additional restrictive assumptions it may happen that the compressed resolvent is the resolvent of a dissipative operator. Equivalently, the reduced evolution $V(t)$ forms a strongly continuous semigroup of contractions,

$$V(t + s) = V(t)V(s), \quad s, t \geq 0,$$

and then a purely exponential decay in such systems is possible. However, in this case, the Hamiltonian of the system necessarily has absolutely continuous spectrum filling in the entire real axis. In particular, the Hamiltonian H is neither semi-bounded from below nor from above.

If we assume that the quantum system is subject to continuous monitoring, the semigroup property for the reduced evolution $V(t)$ is by no means necessary for exponential decay to occur. In some cases it is sufficient to require the existence of the Zeno limit evolution $T(t)$ given by (15.22). Under these more relaxed assumptions one can get rid of the necessity for the Hamiltonian to have a portion of the absolutely continuous spectrum but nevertheless the energy spectrum of the exponentially decaying state must still be unbounded from both sides.

It is worth mentioning that the sufficient conditions discussed above that guarantee the exponential decay of some states of the system i) are by no means necessary and ii) relate to quantities that are not observable (for instance, $U(t)$, $V(t)$ or $T(t)$ are not self-adjoint operators) and therefore the study of the corresponding contractive semi-groups is of mathematical interest only: the natural physical requirement that the Hamiltonian of a quantum system has to a be a semi-bounded operator is violated.

However, as we have shown in Example 15.14, the 1-stable limit theorem suggests a natural sufficient condition for the exponential decay of a state to occur. This condition does not require any of the above hypotheses: the Hamiltonian of the system is positive and has discrete spectrum. In particular, the Zeno limit evolution $T(t)$ is ill defined in this case.

Chapter 16

THE QUANTUM ZENO VERSUS ANTI-ZENO EFFECT ALTERNATIVE

In this chapter we focus our attention on continuous monitoring of massive one-dimensional particles on a semi-axis. We assume that the Hamiltonian for a particle with one degree of freedom is the one-dimensional Schrödinger operator

$$H = -\frac{\hbar^2}{2m}\frac{d^2}{dx^2}$$

in the Hilbert space $L^2((0, \infty))$. In the system of units where $\hbar = 1$ and mass $m = 1/2$ the Hamiltonian is given by the differential expression

$$H = -\frac{d^2}{dx^2}$$

with appropriate boundary conditions at the origin. It turns out that the results of frequent measurements for such quantum systems depend on the specific choice of the boundary conditions at the origin and they differ qualitatively. For instance, in the case of the Dirichlet Schrödinger operator, any smooth initial state with $\phi(0) \neq 0$ is an Anti-Zeno state under the continuous monitoring. In contrast to this, if the quantum evolution is governed by any other self-adjoint realization of the second order differentiation operator, then all smooth initial states are Zeno states.

The proper understanding of this phenomenon requires a more thorough analysis of the decay properties of quantum systems, which we will proceed below.

We start with a definition of a resonant state under continuous monitoring in a non-linear time-scale.

Definition 16.1. Let $\mathfrak{t}(t)$ be an increasing continuous function of t such that

$$\mathfrak{t}(0) = 0.$$

We say that the state ϕ is a resonant state under continuous monitoring of the quantum unitary evolution $\phi \to e^{itH}\phi$ in the time-scale $\mathfrak{t}(t)$ if

$$\lim_{n\to\infty} [p(\mathfrak{t}(t/n))]^n = e^{-\sigma|t|}.$$

where

$$p(t) = |(e^{itH}\phi, \phi)|^2$$

is the survival probability.

Remark 16.2. If $\phi \in \text{Dom}(H)$, then ϕ is a Zeno state by Proposition 15.3 in the standard (linear) time-scale

$$\mathfrak{t}(t) = t,$$

but ϕ is simultaneously a resonant state in the non-linear time-scale

$$\mathfrak{t}(t) = \sqrt{t}$$

as it follows from (15.18) (see Subsection 15.4 where the concept of a prolonged frequent measurement is discussed).

First, we treat the case of the Schrödinger operator on the positive semi-axis with the Dirichlet boundary condition at the origin.

Before formulating the corresponding result recall (see [33]) that if $\mathcal{N}$ denotes the normal distribution

$$\mathcal{N}(\lambda) = \frac{1}{\sqrt{2\pi}} \int_{\infty}^{\lambda} e^{-\frac{1}{2}y^2} dy,$$

then

$$F_\sigma(\lambda) = 2\left[1 - \mathcal{N}\left(\sqrt{\frac{\sigma}{\lambda}}\right)\right], \quad \lambda > 0, \tag{16.1}$$

defines one-sided stable (Lévy) distribution with index of stability $\frac{1}{2}$ the characteristic function $f(t)$ of which is given by

$$f(t) = \exp\left(-\sigma|t|^{1/2}\left(1 - i\frac{t}{|t|}\right)\right). \tag{16.2}$$

It is remarkable that along with the Gaussian and Cauchy distributions, the probability density function of the Lévy distribution is known in closed form [33]

$$\rho(\lambda) = \left(\frac{\sigma}{2\pi}\right)^{1/2} \frac{1}{\lambda^{3/2}} e^{-\frac{\sigma}{2\lambda}}, \quad \lambda > 0.$$

Our first result shows that for a typical initial state ϕ ($\phi \in W_2^2((0, \infty))$) such that $\phi(0) \neq 0$) the spectral measure $(E_H(d\lambda)\phi, \phi)$ of the state ϕ has r-moments for all $r < 1/2$ but not for $r = 1/2$. Here H denotes the Schrödinger operator with the Dirichlet boundary condition at the origin. In particular this means that such states are anti-Zeno states under continuous monitoring of the quantum evolution $\phi \mapsto e^{itH}\phi$. However, in the time scale $t(t) = t^2$ such states do exhibit exponential decay. Equivalently, short frequent measurements yield

$$\lim_{n\to\infty} \left|(e^{it/nH}\phi, \phi)\right|^{2\sqrt{n}} = e^{-\sigma|t|^{1/2}}$$

for some $\sigma > 0$.

Theorem 16.3. *Let $\mathcal{H} = L^2((0, \infty))$ and*

$$H = -\frac{d^2}{dx^2} \tag{16.3}$$

be the Schrödinger operator with the Dirichlet boundary condition at the origin

$$\mathrm{Dom}(H) = \{f \in W_2^2((0, \infty)) \mid f(0) = 0\}.$$

Suppose that $\phi \in W_2^2((0, \infty))$ is such that $\phi(0) \neq 0$ and $\|\phi\| = 1$. Then the distribution function $N(\lambda)$ of the spectral measure

$$\nu_\phi(d\lambda) = (E_H(d\lambda)\phi, \phi)$$

of the element ϕ belongs to the domain of normal attraction of the one-sided $\frac{1}{2}$-stable Lévy distribution F_σ (16.1) the characteristic function of which is given by (16.2) with

$$\sigma = \sqrt{\frac{2}{\pi}}|\phi(0)|^2.$$

In particular, the state ϕ is a resonant state in the time-scale

$$t(t) = t^2.$$

That is,

$$\lim_{n\to\infty} [p(t(t/n))]^n = \exp(-2\sigma|t|), \tag{16.4}$$

where

$$p(t) = |(e^{itH}\phi, \phi)|^2$$

is the survival probability.

Proof. Let $\dot{H}$ be the restriction of H on

$$\mathrm{Dom}(\dot{H}) = \{f \in \mathrm{Dom}(H) \mid f(0) = f'(0) = 0\}.$$

It is known that the Weyl-Titchmarsh function $M(z)$ associated with the pair $(\dot{H}, H)$ admits the representation [42]

$$M(z) = i\sqrt{2z} + 1, \quad z \in \mathbb{C}_+.$$

By the Stieltjes inversion formula, we have that

$$M(z) = \int_0^\infty \left(\frac{1}{\lambda - z} - \frac{\lambda}{\lambda^2 + 1} \right) d\mu(\lambda),$$

where $\mu(d\lambda)$ is an absolutely continuous measure supported by the positive semi-axis with the density

$$\frac{d\mu(\lambda)}{d\lambda} = \frac{1}{\pi}\mathrm{Im}(M(\lambda + i0)) = \frac{\sqrt{2}}{\pi}\sqrt{\lambda}, \quad \lambda > 0. \tag{16.5}$$

Suppose that $g_\pm$, $\|g_\pm\| = 1$, are deficiency elements $g_\pm \in \mathrm{Ker}((\dot{H})^* \mp iI)$ such that

$$g_+ - g_- \in \mathrm{Dom}(H). \tag{16.6}$$

In fact, the deficiency elements $g_\pm$ of the symmetric operator $\dot{H}$ can be chosen as (see, e.g., [42])

$$g_+(x) = 2^{1/4}e^{i\frac{\sqrt{2}}{2}x}e^{-\frac{\sqrt{2}}{2}x} \quad \text{and} \quad g_-(x) = \overline{g_+(x)}, \quad x \geq 0. \tag{16.7}$$

In this case, $g_+(0) - g_-(0) = 0$ which shows that (16.6) holds.

One can apply Theorem C.1 in Appendix C to conclude that there is a unitary map U from $L^2((0, \infty))$ onto $L^2((0, \infty); d\mu)$ such that UHU^{-1} is the operator of multiplication by independent variable in $L^2((0, \infty); d\mu)$.

Since (16.6) holds, from Remark C.2 in Appendix C it follows that

$$(Ug_\pm)(\lambda) = \frac{\Theta}{\lambda \mp i}, \quad \lambda > 0, \tag{16.8}$$

for some $|\Theta| = 1$.

By the hypothesis, $\phi \in W_2^2((0, \infty)) = \text{Dom}((\dot{H})^*)$. Therefore, in accordance with von Neumann's formula the element ϕ admits the representation

$$\phi = \alpha g_+ + \beta g_- + h \tag{16.9}$$

for some uniquely determined $\alpha, \beta \in \mathbb{C}$ and $h \in \text{Dom}(\dot{H}) \subset \text{Dom}(H)$.

We claim that the distribution function $N(\lambda)$ of the spectral measure

$$\nu_\phi(d\lambda) = (E_H(d\lambda)\phi, \phi)$$

of the element ϕ admits the asymptotic representation

$$1 - N(\lambda) = \frac{2\sqrt{2}}{\pi} \frac{|\alpha + \beta|^2}{\sqrt{\lambda}} + o(\lambda^{-5/4}) \quad \text{as } \lambda \to \infty. \tag{16.10}$$

Indeed, from (16.8) and (16.9) it follows that

$$(U\phi)(\lambda) = \frac{a}{\lambda - i} + \frac{b}{\lambda + i} + (Uh)(\lambda), \tag{16.11}$$

where $a = \Theta\alpha$ and $b = \Theta\beta$. In particular,

$$|a + b| = |\alpha + \beta|. \tag{16.12}$$

Working out the computations in the model representation provided by Theorem C.1 in Appendix C, we obtain for the distribution function $N(\lambda)$ the representation

$$1 - N(\lambda) = \int_\lambda^\infty \left| \frac{a}{s - i} + \frac{b}{s + i} + (Uh)(s) \right|^2 d\mu(s) = |a + b|^2 \int_\lambda^\infty \frac{d\mu(s)}{s^2 + 1}$$

$$+ 2\text{Re}\, a\bar{b} \int_\lambda^\infty \left(\frac{1}{(s - i)^2} - \frac{1}{s^2 + 1} \right) d\mu(s)$$

$$+ 2\text{Re} \int_\lambda^\infty \left(\frac{a}{s - i} + \frac{b}{s + i} \right) \overline{(Uh)(s)} d\mu(s)$$

$$+ \int_\lambda^\infty |(Uh)(s)|^2 d\mu(\lambda), \quad \lambda \geq 0. \tag{16.13}$$

From (16.5) it follows

$$\int_\lambda^\infty \frac{d\mu(s)}{s^2 + 1} = \frac{\sqrt{2}}{\pi} \int_\lambda^\infty \frac{\sqrt{s}}{s^2 + 1} ds = \frac{2\sqrt{2}}{\pi} \frac{1}{\sqrt{\lambda}} + O(\lambda^{-3/2}) \quad \text{as } \lambda \to \infty. \tag{16.14}$$

Therefore, for the first term of the right hand side of (16.13) we have the asymptotic representation

$$|a + b|^2 \int_\lambda^\infty \frac{d\mu(s)}{s^2 + 1} = |\alpha + \beta|^2 \frac{2\sqrt{2}}{\pi} \frac{1}{\sqrt{\lambda}} + O(\lambda^{-3/2}).$$

Here we have used (16.12).

The remaining three terms in (16.13) can be estimated as follows

$$\int_\lambda^\infty \left| \frac{1}{(s - i)^2} - \frac{1}{s^2 + 1} \right| d\mu(s) \leq \frac{3}{\lambda} \int_\lambda^\infty \frac{d\mu(s)}{s^2 + 1} = O(\lambda^{-3/2}), \qquad (16.15)$$

$$\left| \int_\lambda^\infty \frac{\overline{(Uh)(s)}}{s \pm i} d\mu(s) \right|$$

$$\leq \frac{1}{\lambda} \sqrt{\int_\lambda^\infty \frac{d\mu(s)}{s^2 + 1}} \cdot \sqrt{\int_\lambda^\infty (1 + s^2) |(Uh)(s)|^2 d\mu(s)}$$

$$= o(\lambda^{-5/4}), \qquad (16.16)$$

and

$$\int_\lambda^\infty |(Uh)(s)|^2 \, d\mu(s) \leq \frac{1}{\lambda^2} \int_\lambda^\infty (1 + s^2) |(Uh)(s)|^2 d\mu(s) = o(\lambda^{-2}),$$

$$\text{as } \lambda \to \infty. \qquad (16.17)$$

Here, in (16.16) and (16.17) we have used that $h \in \text{Dom}(\dot{H}) \subset \text{Dom}(H)$, so that

$$Uh \in L^2(\mathbb{R}; (1 + \lambda^2) d\mu(\lambda)).$$

Combining (16.14) and the asymptotic estimates (16.15)-(16.17), from (16.13), we get

$$1 - N(\lambda) = \frac{2\sqrt{2}}{\pi} \frac{|\alpha + \beta|^2}{\sqrt{\lambda}} + o(\lambda^{-5/4}) \quad \text{as } \lambda \to \infty. \qquad (16.18)$$

Next, we evaluate $|\alpha + \beta|^2$ via the boundary data $|\phi(0)|^2$.

One observes (see (16.9)) that the boundary condition

$$\phi(0) = \alpha g_+(0) + \beta g_-(0) + h(0)$$

holds. Now, since $h \in \text{Dom}(\dot{H})$, we have $h(0) = 0$, so that

$$\phi(0) = \alpha g_+(0) + \beta g_-(0) = 2^{1/4}(\alpha + \beta)$$

as it follows from (16.7). Hence,

$$|\alpha + \beta|^2 = \frac{|\phi(0)|^2}{\sqrt{2}}.$$

Now, taking into account that $\phi(0) \neq 0$, from (16.18) we get the asymptotic representation

$$N(\lambda) = 1 - \frac{2}{\pi} \frac{|\phi(0)|^2}{\lambda^{1/2}} (1 + o(1)) \quad \text{as } \lambda \to \infty.$$

Moreover, since H is a non-negative operator, we obviously have

$$N(\lambda) = 0, \quad \lambda < 0.$$

By Theorem H.1 in Appendix H, the distribution $N(\lambda)$ belongs to the domain of normal attraction of the one-sided stable Lévy distribution F_σ with the characteristic function

$$f(t) = \exp\left(-\sigma|t|^{1/2}\left(1 - i\frac{t}{|t|}\right)\right), \tag{16.19}$$

where

$$\sigma = \sqrt{\frac{2}{\pi}}|\phi(0)|^2.$$

Indeed, the distribution $N(\lambda)$ satisfies the conditions (H.4) and (H.5) of Theorem H.1 in Appendix H with $\alpha = \frac{1}{2}$,

$$c_1 = \frac{2}{\pi}|\phi(0)|^2 \quad \text{and} \quad c_2 = 0.$$

Therefore, $N(\lambda)$ belongs to the domain of normal attraction of a stable law with the characteristic function

$$f(t) = \exp\left(-\sigma|t|^{1/2}\left(1 - i\beta\frac{t}{|t|}\omega\left(t, \frac{1}{2}\right)\right)\right).$$

Here

$$\sigma = (c_1 + c_2)d\left(\frac{1}{2}\right) = \frac{2}{\pi}|\phi(0)|^2 \cdot d\left(\frac{1}{2}\right),$$

$$\beta = \frac{c_1 - c_2}{c_1 + c_2} = 1,$$

$$\omega\left(t, \frac{1}{2}\right) = \tan\left(\frac{\pi}{4}\right) = 1,$$

and

$$d\left(\frac{1}{2}\right) = \Gamma(1/2)\cos\frac{\pi}{4} = \frac{1}{2}\sqrt{2\pi}.$$

The main assertion of the theorem is now proven.

To complete the proof of the theorem it remains to apply the $1/2$-stable limit theorem to see that

$$\lim_{n\to\infty}\left|\left(\exp\left(in^{-2}tH\right)\phi,\phi\right)\right|^{2n} = \exp\left(-2\sigma|t|^{1/2}\right), \tag{16.20}$$

which justifies (16.4) by a change of variables. $\qquad\square$

The situation is quite different for any other self-adjoint realization of the free Schrödinger operator H' on the semi-axis. In this case, the spectral measure $(E_H(d\lambda)\phi,\phi)$ of a typical state $\phi \in W_2^2((0,\infty))$ has r-moments for all $r < 3/2$. As a consequence, such states are Zeno states under continuous monitoring of the quantum evolution $\phi \mapsto e^{itH'}\phi$. However, ϕ becomes a resonant state in the time scale $\mathfrak{t}(t) = t^{2/3}$. Equivalently, prolonged measurements "unfreeze" the quantum system and

$$\lim_{n\to\infty}\left|(e^{it/nH}\phi,\phi)\right|^{2n\sqrt{n}} = e^{-\sigma'|t|^{3/2}}$$

for some $\sigma' > 0$.

Theorem 16.4. *Let* $\mathcal{H} = L^2((0,\infty))$, $\gamma \in \mathbb{R}$ *and*

$$H' = -\frac{d^2}{dx^2} \tag{16.21}$$

be the Schrödinger operator with the mixed boundary condition at the origin

$$\mathrm{Dom}(H') = \{f \in W_2^2((0,\infty)) \mid f'(0) + \gamma f(0) = 0\}.$$

Suppose that $\phi \in W_2^2((0,\infty))$, $\|\phi\| = 1$, *and assume, in addition, that*

$$\phi'(0) + \gamma\phi(0) \neq 0.$$

Then the distribution function $N(\lambda)$ *of the spectral measure*

$$\nu_\phi(d\lambda) = (E_{H'}(d\lambda)\phi,\phi)$$

of the element ϕ *belongs to the domain of normal attraction of the* $3/2$-*stable law with the characteristic function*

$$f(t) = \exp\left(-\sigma'|t|^{3/2}(1 + i\,\mathrm{sgn}(t))\right),$$

where

$$\sigma' = \frac{2}{3}\sqrt{\frac{2}{\pi}}|\phi'(0) + \gamma\phi(0)|^2.$$

In particular, the state ϕ is a resonant state in the time-scale

$$\mathfrak{t}(t) = t^{2/3}.$$

That is,

$$\lim_{n \to \infty} [p(\mathfrak{t}(t/n))]^n = \exp(-2\sigma'|t|), \tag{16.22}$$

where

$$p(t) = |(e^{itH}\phi, \phi)|^2$$

is the survival probability.

Proof. Let $\dot{H}$ be the symmetric restriction of the operator H' on

$$\text{Dom}(\dot{H}) = \{f \in \text{Dom}(H') \mid f(0) = f'(0) = 0\}.$$

Denote by $g_{\pm}$, $\|g_{\pm}\| = 1$, the deficiency elements of the symmetric operator $\dot{H}$

$$g_+(x) = 2^{1/4}e^{i\frac{\sqrt{2}}{2}x}e^{-\frac{\sqrt{2}}{2}x} \quad \text{and} \quad g_-(x) = \overline{g_+(x)}\Theta, \tag{16.23}$$

where Θ is chosen in such a way to ensure that

$$g_+ - g_- \in \text{Dom}(H'). \tag{16.24}$$

The parameter Θ can be determined as follows. From (16.24) it follows that $g_+(x) - g_-(x)$ should satisfy the boundary condition

$$(g'_+(0) - g'_-(0)) + \gamma(g_+(0) - g_-(0)) = 0$$

and hence

$$(\zeta - \bar{\zeta}\Theta) + \gamma(1 - \Theta) = 0, \tag{16.25}$$

where

$$\zeta = \frac{\sqrt{2}}{2}(1 - i). \tag{16.26}$$

Solving (16.25) for Θ yields

$$\Theta = \frac{\zeta + \gamma}{\bar{\zeta} + \gamma}.$$

Since (16.24) holds, one can apply Theorem C.1 in Appendix C, and the same reasoning as the one in the proof of Theorem 16.3 shows that the leading term of the asymptotics of the distribution function $N(\lambda)$ is given by

$$1 - N(\lambda) = |\alpha + \beta|^2 \int_\lambda^\infty \frac{d\mu(s)}{s^2 + 1} + o(\lambda^{-7/4}) \quad \text{as } \lambda \to \infty. \tag{16.27}$$

Here $\mu(d\lambda)$ is the measure associated with the Herglotz-Nevanlinna decomposition for the Weyl-Titchmarsh function $M(z)$ associated with the pair $(\dot{H}, H')$

$$M(z) = \int_{\text{spec}(H')} \left(\frac{1}{\lambda - z} - \frac{\lambda}{\lambda^2 + 1} \right) d\mu(\lambda)$$

and α and β are determined by the von Neumann decomposition

$$\phi = \alpha g_+ + \beta g_- + h, \quad h \in \text{Dom}(\dot{H}). \tag{16.28}$$

To justify the asymptotic representation (16.27) we argue as follows.

First recall, that it is known that the Weyl-Titchmarsh function associated with the pair $(\dot{H}, H')$ has the form

$$M(z) = \frac{\cos \alpha + \sin \alpha (i\sqrt{2z} + 1)}{\sin \alpha - \cos \alpha (i\sqrt{2z} + 1)}, \quad z \in \mathbb{C}_+. \tag{16.29}$$

Here the boundary condition parameter γ and (the von Neumann extension) parameter α are related as [42]

$$\gamma = 2^{-1/2}(1 - \tan \alpha), \quad \alpha \neq \frac{\pi}{2}. \tag{16.30}$$

From (16.29) it follows that the restriction of the measure $\mu(d\lambda)$ on the positive semi-axis is an absolutely continuous measure with the density given by

$$\frac{d\mu(\lambda)}{d\lambda} = \frac{1}{\pi} \text{Im}(M(\lambda + i0)) d\lambda, \quad \lambda > 0.$$

Explicit computations show that

$$\frac{1}{\pi}\mathrm{Im}(M(\lambda+i0))d\lambda$$

$$= \frac{1}{\pi}\mathrm{Im}\frac{\cos\alpha+\sin\alpha(i\sqrt{2\lambda}+1)}{\sin\alpha-\cos\alpha(i\sqrt{2\lambda}+1)}$$

$$= \frac{1}{\pi}\mathrm{Im}\frac{(\cos\alpha+\sin\alpha(i\sqrt{2\lambda}+1))(\sin\alpha-\cos\alpha(-i\sqrt{2\lambda}+1))}{(\sin\alpha-\cos\alpha)^2+2\lambda\cos^2\alpha}$$

$$= \frac{1}{\pi}\mathrm{Im}\frac{(\cos\alpha+\sin\alpha+i\sqrt{2\lambda}\sin\alpha)(\sin\alpha-\cos\alpha+i\sqrt{2\lambda}\cos\alpha)}{(\sin\alpha-\cos\alpha)^2+2\lambda\cos^2\alpha}$$

$$= \frac{1}{\pi}\frac{\sqrt{2\lambda}}{(\sin\alpha-\cos\alpha)^2+2\lambda\cos^2\alpha}, \quad \lambda>0.$$

Therefore,

$$d\mu(\lambda) = \frac{1}{\pi}\frac{\sqrt{2\lambda}}{(\sin\alpha-\cos\alpha)^2+2\lambda\cos^2\alpha}d\lambda, \quad \lambda>0. \tag{16.31}$$

To justify (16.27), in particular, to see that the error term is of the order of $o(\lambda^{-7/4})$ as $\lambda\to\infty$, we argue exactly as in the proof of Theorem 16.3. To do so, we need to estimate the following three integrals (we use the notation from the proof of Theorem 16.3)

$$I = \int_\lambda^\infty \left|\frac{1}{(s-i)^2}-\frac{1}{s^2+1}\right|d\mu(s),$$

$$II = \left|\int_\lambda^\infty \frac{\overline{(Uh)(s)}}{s\pm i}d\mu(s)\right|,$$

and

$$III = \int_\lambda^\infty |(Uh)(s)|^2\,d\mu(s).$$

We have (as $\lambda\to\infty$)

$$I \le \frac{3}{\lambda}\int_\lambda^\infty \frac{d\mu(s)}{s^2+1} = O(\lambda^{-5/2}). \tag{16.32}$$

Since $h \in \text{Dom}(\dot{H}) \subset \text{Dom}(H)$, and therefore $Uh \in L^2(\mathbb{R}; (1 + \lambda^2)d\mu(\lambda))$, we also have the asymptotic estimates

$$II \leq \frac{1}{\lambda} \sqrt{\int_\lambda^\infty \frac{d\mu(s)}{s^2 + 1}} \cdot \sqrt{\int_\lambda^\infty (1 + s^2)|(Uh)(s)|^2 d\mu(s)} = o(\lambda^{-7/4}) \quad (16.33)$$

and

$$III \leq \frac{1}{\lambda^2} \int_\lambda^\infty (1 + s^2)|(Uh)(s)|^2 d\mu(s) = o(\lambda^{-2}) \quad \text{as } \lambda \to \infty. \quad (16.34)$$

Therefore,

$$I + II + III = o(\lambda^{-7/4}) \quad \text{as } \lambda \to \infty,$$

which completes the justification of the representation (16.27).

Next, combining (16.27) and (16.31) we obtain

$$1 - N(\lambda) = |\alpha + \beta|^2 \int_\lambda^\infty \frac{1}{\pi} \frac{\sqrt{2s}}{(\sin \alpha - \cos \alpha)^2 + 2s \cos^2 \alpha} \frac{ds}{s^2 + 1} + o(\lambda^{-7/4})$$

$$= |\alpha + \beta|^2 \frac{\sqrt{2}}{3\pi \cos^2 \alpha} \lambda^{-3/2}(1 + o(1)) + o(\lambda^{-7/4})$$

$$= |\alpha + \beta|^2 \frac{\sqrt{2}}{3\pi}((\sqrt{2}\gamma - 1)^2 + 1)\lambda^{-3/2}(1 + o(1)) + o(\lambda^{-7/4})$$

$$\text{as } \lambda \to \infty. \quad (16.35)$$

Here we have used the relation

$$\frac{1}{\cos^2 \alpha} = ((\sqrt{2}\gamma - 1)^2 + 1)$$

that easily follows from (16.30). Recall that $\alpha \neq \frac{\pi}{2}$ and therefore $\cos \alpha \neq 0$.

Our next claim is that

$$|\alpha + \beta|^2 = \frac{1}{\sqrt{2}} \frac{1}{(\gamma - \frac{\sqrt{2}}{2})^2 + \frac{1}{2}} |\phi'(0) + \gamma\phi(0)|^2. \quad (16.36)$$

From (16.28) it follows that

$$\phi(0) = \alpha g_+(0) + \beta g_-(0) = \alpha g_+(0) + \beta \Theta \overline{g_+(0)} = (\alpha + \beta \Theta)2^{1/4} \quad (16.37)$$

and

$$\phi'(0) = \alpha g'_+(0) + \beta g'_-(0) = (\alpha\zeta + \beta\overline{\zeta}\Theta)2^{1/4}, \quad (16.38)$$

where ζ is given by (16.26).

Rewriting (16.37) and (16.38) as

$$\begin{pmatrix} 1 & \Theta \\ \zeta & \bar{\zeta}\Theta \end{pmatrix} \begin{pmatrix} \alpha \\ \beta \end{pmatrix} = 2^{-1/4} \begin{pmatrix} \phi(0) \\ \phi'(0) \end{pmatrix},$$

and solving this system of algebraic equations one obtains

$$\begin{pmatrix} \alpha \\ \beta \end{pmatrix} = \frac{1}{\Theta(\bar{\zeta} - \zeta)} \begin{pmatrix} \bar{\zeta}\Theta & -\Theta \\ -\zeta & 1 \end{pmatrix} 2^{-1/4} \begin{pmatrix} \phi(0) \\ \phi'(0) \end{pmatrix}.$$

Therefore,

$$\alpha + \beta = \frac{1}{2^{1/4}\Theta(\bar{\zeta} - \zeta)} [(\bar{\zeta}\Theta - \zeta)\phi(0) + (1 - \Theta)\phi'(0)]$$

$$= \frac{1 - \Theta}{2^{1/4}\Theta(\bar{\zeta} - \zeta)} \left[\frac{\bar{\zeta}\Theta - \zeta}{1 - \Theta}\phi(0) + \phi'(0) \right].$$

From (16.25) it follows that

$$\gamma = \frac{\bar{\zeta}\Theta - \zeta}{1 - \Theta},$$

so that

$$\alpha + \beta = \frac{1 - \Theta}{2^{1/4}\Theta(\bar{\zeta} - \zeta)} [\gamma\phi(0) + \phi(0)].$$

One also observes that

$$\frac{1 - \Theta}{\Theta(\bar{\zeta} - \zeta)} = \frac{1}{\zeta + \gamma},$$

which yields

$$|\alpha + \beta|^2 = \frac{1}{\sqrt{2}} \frac{1}{|\zeta + \gamma|^2} |\phi'(0) + \gamma\phi(0)|^2$$

$$= \frac{1}{\sqrt{2}} \frac{1}{(\gamma - \frac{\sqrt{2}}{2})^2 + \frac{1}{2}} |\phi'(0) + \gamma\phi(0)|^2,$$

and the claim (16.36) follows.

Combining (16.35) and (16.36) and taking into account that $\phi'(0) + \gamma\phi(0) \neq 0$ (by the hypothesis), we finally obtain the asymptotic representation

$$1 - N(\lambda) = |\alpha + \beta|^2 \frac{\sqrt{2}}{3\pi} ((\sqrt{2}\gamma - 1)^2 + 1)\lambda^{-3/2}(1 + o(1))$$

$$= \frac{1}{\sqrt{2}} \cdot \frac{\sqrt{2}}{3\pi} \frac{(\sqrt{2}\gamma - 1)^2 + 1}{(\gamma - \frac{\sqrt{2}}{2})^2 + \frac{1}{2}} |\phi'(0) + \gamma\phi(0)|^2 \lambda^{-3/2}(1 + o(1))$$

$$= \frac{2}{3\pi} |\phi'(0) + \gamma\phi(0)|^2 \lambda^{-3/2}(1 + o(1)) \quad \text{as } \lambda \to \infty. \quad (16.39)$$

Notice that for $\gamma < 0$ the operator H' has a simple eigenvalue $\lambda_0 = -\gamma^2$ and therefore $N(\lambda) = 0$ whenever $\lambda < -\gamma^2$, and $N(\lambda) = 0$ for all $\lambda < 0$ if $\gamma \geq 0$. Therefore,

$$\lim_{\lambda \to -\infty} |\lambda|^{3/2} N(\lambda) = 0. \tag{16.40}$$

By Theorem H.1 in Appendix H, the distribution $N(\lambda)$ belongs to the domain of normal attraction of the one-sided $\frac{3}{2}$-stable distribution with the characteristic function

$$f(t) = \exp\left(-\sigma'|t|^{3/2}\left(1 + i\frac{t}{|t|}\right)\right), \tag{16.41}$$

where

$$\sigma' = \frac{2}{3}\sqrt{\frac{2}{\pi}}|\phi'(0) + \gamma\phi(0)|^2.$$

Indeed, the distribution $N(\lambda)$ satisfies the conditions (H.4) and (H.5) of Theorem H.1 in Appendix H with $\alpha = \frac{3}{2}$,

$$c_1 = \frac{2}{3\pi}|\phi'(0) + \gamma\phi(0)|^2 \quad \text{and} \quad c_2 = 0.$$

By Theorem H.1, $N(\lambda)$ belongs to the domain of normal attraction of a stable law with the characteristic function

$$f(t) = \exp\left(-\sigma'|t|^{3/2}\left(1 - i\beta\frac{t}{|t|}\omega\left(t, \frac{3}{2}\right)\right)\right),$$

where the parameters σ' and β are given by

$$\sigma' = (c_1 + c_2)d\left(\frac{3}{2}\right) = \frac{2}{\pi}|\phi'(0) + \gamma\phi(0)|^2 \cdot d\left(\frac{3}{2}\right),$$

$$\beta = \frac{c_1 - c_2}{c_1 + c_2} = 1,$$

with

$$d\left(\frac{3}{2}\right) = \Gamma(-1/2)\cos\frac{3\pi}{4} = \frac{1}{2}\sqrt{2\pi},$$

and

$$\omega\left(t, \frac{3}{2}\right) = \tan\left(\frac{3\pi}{4}\right) = -1.$$

Now (16.41) follows.

To complete the proof of the theorem it remains to apply the 3/2-stable limit theorem to see that

$$\lim_{n \to \infty} \left| \left(\exp\left(in^{-2/3} tH \right) \phi, \phi \right) \right|^{2n} = \exp\left(-2\sigma' |t|^{3/2} \right),$$ (16.42)

which justifies (16.22) by a change of variables. □

Remark 16.5. The right hand side of (16.42) is the characteristic functions of the Holtsmark distribution [47]. The Holtsmark distribution is a special case of a symmetric stable distribution with the index of stability $\alpha = 3/2$ and skewness parameter $\beta = 0$ (see Appendix H, eqs. (H.1), (H.2) with $\alpha = 3/2$ and $\beta = \gamma = 0$).

Scholium. The 1/2- and 3/2-stable limit theorems, Theorems 16.3 and 16.4, respectively, show that the results of continuous monitoring of the quantum evolution of a smooth state ϕ are rather sensitive to the choice of a self-adjoint realization of the Hamiltonian, the Schrödinger operator (16.3) and (16.21), respectively.

For instance, for the Schrödinger operator H with the Dirichlet boundary condition at the origin we have

$$\lim_{n \to \infty} |(e^{it/nH} \phi, \phi)|^{2\sqrt{n}} = e^{-2\sigma |t|^{1/2}},$$

where

$$\sigma = \sqrt{\frac{2}{\pi}} |\phi(0)|^2.$$

Therefore, if the probability density $|\phi(0)|^2$ to find a quantum particle at the origin does not vanish, then the state ϕ is an anti-Zeno state. That is,

$$\lim_{n \to \infty} |(e^{it/nH} \phi, \phi)|^{2n} = 0.$$

In the meanwhile, for the Schrödinger operator H' with the mixed boundary condition

$$f'(0) + \gamma f(0) = 0,$$

one obtains that

$$\lim_{n \to \infty} |(e^{it/nH'} \phi, \phi)|^{2n\sqrt{n}} = e^{-2\sigma' |t|^{3/2}},$$

where

$$\sigma' = \frac{2}{3} \sqrt{\frac{2}{\pi}} |\phi'(0) + \gamma \phi(0)|^2.$$

Hence,

$$\lim_{n \to \infty} |(e^{it/nH'}\phi, \phi)|^{2n} = 1.$$

In other words, any smooth state ϕ is a Zeno state under the continuous monitoring of the evolution $\phi \mapsto e^{it/nH'}\phi$ where H' is any self-adjoint realizations of the second differentiation operator different form the Friedrichs extension H of $\dot{H}$.

We summarize the observations above in a more formal way.

Corollary 16.6. *Suppose that $\phi \in W_2^2((0, \infty))$, $\|\phi\| = 1$.*

(i) *Let H be the Schrödinger operator with the Dirichlet boundary condition at the origin. Then ϕ is a Zeno state under the continuous monitoring of the unitary evolution $\phi \mapsto e^{itH}\phi$ if and only if $\phi(0) = 0$. Otherwise, ϕ is an anti-Zeno state.*

(ii) *If H' is any other self-adjoint realization of the differential expression*

$$\tau = -\frac{d^2}{dx^2}$$

different from its Friedrichs extension, then ϕ is a Zeno state under the continuous monitoring of the unitary evolution $\phi \mapsto e^{itH'}\phi$.

Proof. (i) If $\phi(0) = 0$, then $\phi \in \text{Dom}(H)$ and therefore ϕ is a Zeno state under the continuous monitoring of the unitary evolution $\phi \mapsto e^{itH}\phi$. If $\phi(0) \neq 0$, by Theorem 16.3 the distribution function of the spectral measure of the element ϕ belongs to the domain of attraction of a $1/2$-stable law, and therefore ϕ is an anti-Zeno state by Lemma 15.7.

(ii) Notice that for any self-adjoint extension H' different from the Friedrichs extension H the domain of the quadratic form of H' coincides with the Sobolev class $W_2^1((0, \infty))$. Since $\phi \in W_2^2((0, \infty)) \subset W_2^1((0, \infty))$, we have that the distribution $N(\lambda)$ of the spectral measure $(\mathsf{E}_{H'}(d\lambda)\phi, \phi)$ of the state ϕ has the first moment and hence ϕ is necessarily a Zeno state. $\square$

Remark 16.7. (i) In the case of the Schrödinger operator H with the Dirichlet boundary condition at the origin, one can slightly relax the smoothness requirement on the state ϕ that $\phi \in W_2^2((0, \infty))$: If $\phi \in W_2^1((0, \infty))$ only and $\phi(0) = 0$, then the state ϕ belongs to the domain of the quadratic form of the Schrödinger operator H. In this case, ϕ is also a Zeno state under the continuous monitoring of the unitary evolution $\phi \mapsto e^{itH}\phi$ by Proposition 15.3.

(ii) From Theorem 16.4 it follows that the spectral measure

$$\nu_\phi(d\lambda) = (\mathsf{E}_{H'}(d\lambda)\phi, \phi)$$

of the state ϕ has moments of order r for all $r < \frac{3}{2}$. In particular, the state ϕ belongs to the domain of the quadratic form of H' and therefore, ϕ is a Zeno state under the continuous monitoring of the unitary evolution $\phi \mapsto e^{itH'}\phi$ by Proposition 15.3.

(iii). As far as the domain issues are concerned, we have the following inclusions

$$\mathrm{Dom}(\dot{H}) \subset \mathrm{Dom}(H) \subset \mathrm{Dom}((\dot{H})^*) = W_2^2((0, \infty))$$

and

$$\mathrm{Dom}((\dot{H})^*) \cap \mathrm{Dom}(H^{1/2}) = \mathrm{Dom}(H) = \{f \in W_2^2((0, \infty)) \mid f(0+) = 0\}$$

for the Friedrichs extension H of $\dot{H}$, $H \geq 0$. For any self-adjoint extension H' different from the Friedrichs extension H we have

$$\mathrm{Dom}(\dot{H}) \subset \mathrm{Dom}(H') \subset \mathrm{Dom}((\dot{H})^*) \subset \mathrm{Dom}(|H'|^{1/2}) = W_2^1((0, \infty))$$

and therefore

$$\mathrm{Dom}((\dot{H})^*) \cap \mathrm{Dom}(|H'|^{1/2}) = \mathrm{Dom}((\dot{H})^*) = W_2^2((0, \infty)).$$

Notice that for the Friedrichs extension H we have

$$\mathrm{Dom}(H^{1/2}) = \{f \in W_2^1((0, \infty)) \mid f(0+) = 0\}$$
$$\neq \mathrm{Dom}(|H'|^{1/2}) = W_2^1((0, \infty)),$$

which explains the peculiar "phase transition" in the relative geometry of domains (the Sobolev spaces) when replacing the Friedrichs extension H with any other self-adjoint extension H'.

Chapter 17

THE QUANTUM ZENO EFFECT VERSUS EXPONENTIAL DECAY ALTERNATIVE

Throughout this chapter we assume that $\mathbb{Y}$ is a metric graph in one of the Cases (i)–(iii) (see the classification in the beginning of Chapter 4). Denote by $(\dot{D}, \widehat{D}, D)$ the triple of differentiation operators on $\mathbb{Y}$ as introduced in Chapter 9.

Recall that in Case (i), the metric graph has the form $\mathbb{Y} = (-\infty, 0) \sqcup (0, \infty)$, in Case (ii), $\mathbb{Y} = (0, \ell)$, in Case (iii), $\mathbb{Y} = (-\infty, 0) \sqcup (0, \infty) \sqcup (0, \ell)$. Also recall that the reference self-adjoint operator D is the differentiation operator on the graph $\mathbb{Y}$ defined on

$$\text{Dom}(D) = \{ f_\infty \in W_2^1(\mathbb{Y}) \mid f_\infty(0+) = -f_\infty(0-) \},$$

$$\text{Dom}(D) = \{ f_\ell \in W_2^1(\mathbb{Y}) \mid f_\ell(0) = -f_\ell(\ell) \},$$

$$\text{Dom}(D) = \left\{ f_\infty \oplus f_\ell \in W_2^1(\mathbb{Y}) \, \middle| \, \begin{cases} f_\infty(0+) = k f_\infty(0-) + \sqrt{1 - k^2} f_\ell(\ell) \\ f_\ell(0+) = \sqrt{1 - k^2} f_\infty(0-) - k f_\ell(\ell) \end{cases} \right\},$$

in Cases (i)–(iii), respectively. Here $0 < k < 1$ is the parameter from the boundary condition (4.3) (the quantum gate coefficient) that determines the symmetric operator $\dot{D}$ in Case (iii).

More generally, see Theorem 17.4 below, we will also deal with the triples $(\dot{D}, \widehat{D}, D_\Theta)$ where D_Θ, $|\Theta| = 1$ is the self-adjoint operator referred to in Theorem 5.1.

Our main concern is to study small-time asymptotic behavior of the quantum survival probability

$$p(t) = |(e^{itH}\phi, \phi)|^2 \quad \text{as } t \to 0,$$

where $H = D$, or, more generally, $H = D_\Theta$, the magnetic Hamiltonian. We also assume that the state ϕ belongs to the test space

$$\mathcal{L} = \text{Dom}((\dot{D})^*).$$

We obviously have the inclusion

$$\mathcal{L} \subsetneq \bigoplus_{e \subset \Upsilon} W_2^1(e),$$

where the sum is taken over all edges e of the graph Υ.

The main goal of this chapter is to show that the survival probability under continuous monitoring of the quantum evolution

$$\phi \mapsto e^{itH}\phi, \quad \phi \in \mathcal{L},$$

on the metric graph either experiences an exponential decay or, alternatively, the quantum Zeno effect takes place. This justifies the complementarity of the Exponential Decay and the Quantum Zeno Effect scenarios for hyperbolic systems first indicated in [66].

We start our analysis with the observation that the normalized deficiency elements of the symmetric operator $\dot{D}$ are resonant states under continuous monitoring of the unitary evolution $\phi \mapsto e^{itH}\phi$ where $H = D$.

Lemma 17.1. *Suppose that a metric graph Υ is in one of the Cases (i)–(iii) and $\dot{D}$ is the symmetric differentiation operator on Υ with boundary conditions* (4.1), (4.2) *and* (4.3), *respectively. Let $g_\pm \in \text{Ker}((\dot{D})^* \mp iI)$, $\|g_\pm\| = 1$, be normalized deficiency elements $g_\pm$ of the symmetric operator $\dot{D}$. Then $g_\pm$ are equidistributed, that is, $g_\pm$ have the same spectral measure*

$$\nu(d\lambda) = (\mathsf{E}_H(d\lambda)g_+, g_+) = (\mathsf{E}_H(d\lambda)g_-, g_-), \tag{17.1}$$

where the Hamiltonian H is given by the differentiation operator D.

Moreover,

$$\lim_{\lambda \to +\infty} \lambda \nu\left((\lambda, \infty)\right) = \lim_{\lambda \to +\infty} \lambda \nu\left((-\infty, -\lambda)\right)$$

$$= \frac{1}{\pi} \begin{cases} 1, & \text{in Case (i)} \\ \coth \dfrac{\ell}{2}, & \text{in Case (ii)} \\ \coth \dfrac{\ell + \ell'}{2}, & \text{in Case (iii).} \end{cases} \tag{17.2}$$

Here, in Case (iii),

$$\ell' = \log \frac{1}{k}$$

and $0 < k < 1$ is the quantum gate coefficient from the boundary condition (4.3) that determines the symmetric operator $\dot{D}$ in Case (iii).

In particular, the deficiency elements $g_\pm$ are resonant states with respect to the continuous monitoring of the unitary dynamics $g_\pm \mapsto e^{itH} g_\pm$.

In this case,

$$\lim_{n \to \infty} |(e^{it/nH} g_\pm, g_\pm)|^{2n} = e^{-\tau |t|}, \tag{17.3}$$

where the decay constant τ is given by

$$\tau = 2 \begin{cases} 1, & \text{in Case (i)} \\ \coth \dfrac{\ell}{2}, & \text{in Case (ii)} \\ \coth \dfrac{\ell + \ell'}{2}, & \text{in Case (iii).} \end{cases}$$

Proof. Let $M(z)$ be the Weyl-Titchmarsh function associated with the pair $(\dot{D}, D)$. By Corollary 7.2,

$$M(z) = \int_{\mathbb{R}} \left(\frac{1}{\lambda - z} - \frac{\lambda}{1 + \lambda^2} \right) d\mu(\lambda), \tag{17.4}$$

where the measure $\mu(d\lambda)$ is given by

$$\mu(d\lambda) = \frac{1}{\pi} \begin{cases} d\lambda, & \text{in Case (i)} \\ \dfrac{2\pi}{\ell} \coth \dfrac{\ell}{2} \sum_{k \in \mathbb{Z}} \delta_{\frac{(2k+1)\pi}{\ell}}(d\lambda), & \text{in Case (ii)} \\ \coth \dfrac{\ell + \ell'}{2} P_{e^{-\ell'}}(\ell\lambda - \pi)\, d\lambda, & \text{in Case (iii).} \end{cases} \tag{17.5}$$

Here, in Case (iii),

$$P_r(\varphi) = \frac{1 - r^2}{1 + r^2 - 2r \cos \varphi}$$

denotes the Poisson kernel.

Recall that by Lemma 4.4 the operator $\dot{D}$ is a prime symmetric operator. Therefore, Theorem C.1 in Appendix C ensures the existence of a unitary map $\mathcal{U}$ from $L^2(\mathbb{Y})$ onto the Hilbert space $L^2(\mathbb{R}, d\mu)$, where $\mu(d\lambda)$ is given by (17.5), with the following properties:

(i) $\mathcal{U}D\mathcal{U}^{-1}$ coincides with the operator of multiplication by independent variable and

(ii) the deficiency elements $g_\pm$ get mapped to simple fractions

$$(\mathcal{U}g_\pm)(\lambda) = \frac{\Theta_\pm}{\lambda \mp i} \quad \text{for some } |\Theta_\pm| = 1.$$

In particular, for any Borel set $\delta \subset \mathbb{R}$ we have

$$(\mathsf{E}_H(\delta)g_+, g_+) = (\mathsf{E}_H(\delta)g_-, g_-) = \int_\delta \frac{d\mu(s)}{s^2 + 1},$$

which shows that $g_\pm$ are equidistributed and hence the spectral measure $\nu(d\lambda)$ in (17.1) is well defined.

It follows that

$$\lambda\nu((\lambda, \infty)) = \lambda \int_\lambda^\infty \frac{d\mu(s)}{s^2 + 1}.$$

In Case (i), in view of (17.5) we have the following asymptotic representation

$$\lambda\nu((\lambda, \infty)) = \lambda \int_\lambda^\infty \frac{1}{\pi} \frac{ds}{s^2 + 1} = \frac{1}{\pi}(1 + o(1)) \quad \text{as } \lambda \to +\infty,$$

which proves that the first limit in (17.2) exists and coincides with the right hand side of (17.2).

In Case (ii), by (17.5),

$$\lambda\nu((\lambda, \infty)) = \lambda\frac{2}{\ell} \coth \frac{\ell}{2} \sum_{\frac{(2k+1)\pi}{\ell} \geq \lambda} \frac{1}{\left(\frac{(2k+1)\pi}{\ell}\right)^2 + 1}$$

$$= \frac{2}{\ell} \coth \frac{\ell}{2} \left(\frac{\ell}{2\pi}\right)^2 \lambda \int_{\frac{\lambda\ell}{2\pi}}^\infty \frac{dk}{k^2} \cdot (1 + o(1))$$

$$= \frac{2}{\ell} \coth \frac{\ell}{2} \cdot \frac{\ell}{2\pi} (1 + o(1))$$

$$= \frac{1}{\pi} \coth \frac{\ell}{2} (1 + o(1)) \quad \text{as } \lambda \to +\infty,$$

proving the first equality (17.2) in Case (ii).

Finally, in Case (iii), using (17.5) we have

$$\lambda \nu((\lambda, \infty)) = \frac{1}{\pi} \coth \frac{\ell + \ell'}{2} \lambda \int_\lambda^\infty P_{e^{-\ell'}} (\ell s - \pi) \frac{ds}{s^2 + 1}$$

$$= \frac{1}{\pi} \coth \frac{\ell + \ell'}{2} (1 + o(1)) \quad \text{as } \lambda \to +\infty,$$

which shows that the first limit in (17.2) exists and coincides with the right hand side of (17.2). Here we used that the Poisson kernel admits the representation

$$P_r(s) = \frac{1 - r^2}{1 + r^2 - 2r \cos s} = 1 + G_r(s),$$

where $G_r(s)$ is a bounded 2π-periodic function with zero mean over the period such that

$$\lim_{\lambda \to +\infty} \lambda \int_\lambda^\infty G_{e^{-\ell'}} (\ell s - \pi) \frac{ds}{s^2 + 1} = 0.$$

Notice that the equality above can be justified by integration by parts.

In a completely similar way one shows that in all Cases (i)–(iii) the second limit in (17.2) exists and coincides with the right hand side of (17.2). □

Remark 17.2. In Case (i), one can apply the residue theorem to see that the survival probability amplitude $(e^{itD} g_\pm, g_\pm)$ itself is exponentially decaying as

$$(e^{itD} g_\pm, g_\pm) = \frac{1}{\pi} \int_{-\infty}^\infty e^{i\lambda t} \frac{d\lambda}{\lambda^2 + 1} = e^{-|t|},$$

which in particular implies (17.3). In this case the result of continuous monitoring of the corresponding quantum system on the time interval $[0, t]$ and a "one time observation" at the moment of time t are identical. That is,

$$(e^{it/nD} g_\pm, g_\pm)^n = (e^{itD} g_\pm, g_\pm)$$

and therefore

$$|(e^{it/nD}g_\pm, g_\pm)|^{2n} = |(e^{itD}g_\pm, g_\pm)|^2 \quad \text{for all } t.$$

In this exceptional (resonant) case the continuous monitoring can neither stop nor modify the evolution. Notice that in this case the energy distribution of the states $g_\pm$ has a typical Cauchy-Lorentz (Breight-Wigner) shape which yields a purely exponential decay, see [29, Example 1.2.4], cf. [93, p. 759, a counterexample].

To understand better fine decay properties of a particular state from the test space $\mathcal{L} = \mathrm{Dom}(\dot{D}^*)$ we need a comprehensive information about the boundary functionals associated with the von Neumann decomposition of the test space

$$\mathcal{L} = \mathrm{Dom}((\dot{D})^*) = \mathrm{Ker}((\dot{D})^* - iI) \dotplus \mathrm{Ker}((\dot{D})^* + iI) \dotplus \mathrm{Dom}(\dot{D}). \quad (17.6)$$

Lemma 17.3. *Suppose that a metric graph* $\mathbb{Y}$ *is in one of the Cases* (i)–(iii) *and* $\dot{D}$ *is the symmetric differentiation operator on* $\mathbb{Y}$ *with boundary conditions* (4.1), (4.2) *and* (4.3), *respectively. Denote by* $g_\pm$ *the deficiency elements of the symmetric operator* $\dot{D}$ *referred to in Lemma 4.3.*
Assume that $\phi \in \mathcal{L} = \mathrm{Dom}((\dot{D})^*)$ *and let*

$$\phi = \alpha g_+ + \beta g_- + f, \quad \phi \in \mathrm{Dom}((\dot{D})^*), \quad (17.7)$$

be the decomposition associated with von Neumann's formula (17.6), *where* $\alpha, \beta \in \mathbb{C}$ *and* $f \in \mathrm{Dom}(\dot{D})$.
Then

$$\alpha + \beta = \frac{1}{\sqrt{2}} \begin{cases} \phi_\infty(0-) + \phi_\infty(0+), & \text{in Case } (i) \\[2mm] \sqrt{\tanh \dfrac{\ell}{2}} \left(\phi_\ell(0) + \phi_\ell(\ell) \right), & \text{in Case } (ii) \\[2mm] \sqrt{\tanh \dfrac{\ell + \ell'}{2}} \left(\phi_\ell(\ell) - \dfrac{\phi_\infty(0+) - k\phi_\infty(0-)}{\sqrt{1 - k^2}} \right), & \text{in Case } (iii), \end{cases} \quad (17.8)$$

where, in Case (iii),

$$\ell' = \log \frac{1}{k}$$

and $0 < k < 1$ *is the quantum gate coefficient from the boundary condition* (4.3) *that determines the symmetric operator* $\dot{D}$ *in Case* (iii).

Proof. In Case (i), we have

$$\phi_\infty(x) = \alpha\sqrt{2}e^x \chi_{(-\infty,0)}(x) + \beta\sqrt{2}e^{-x}\chi_{(0,\infty)}(x) + f(x)$$

for some $f \in \text{Dom}(\dot{D})$.

Since $f(0-) = f(0+) = 0$, we have

$$\phi_\infty(0-) = \alpha\sqrt{2} \quad \text{and} \quad \phi_\infty(0+) = \beta\sqrt{2}.$$

Therefore

$$\alpha + \beta = \frac{\phi_\infty(0+) + \phi_\infty(0+)}{\sqrt{2}},$$

proving (17.8) in that case.

In Case (ii),

$$\phi_\ell(x) = \alpha\sqrt{\frac{2}{e^{2\ell}-1}}e^x + \beta\sqrt{\frac{2}{e^{2\ell}-1}}e^{\ell-x} + f(x)$$

and therefore

$$\alpha + e^\ell\beta = \sqrt{\frac{e^{2\ell}-1}{2}}\phi_\ell(0)$$

and

$$e^\ell\alpha + \beta = \sqrt{\frac{e^{2\ell}-1}{2}}\phi_\ell(\ell).$$

Hence

$$\alpha + \beta = \sqrt{\frac{e^\ell-1}{e^\ell+1}}\frac{\phi_\ell(0)+\phi_\ell(\ell)}{\sqrt{2}} = \sqrt{\tanh\frac{\ell}{2}} \cdot \frac{\phi_\ell(0)+\phi_\ell(\ell)}{\sqrt{2}},$$

proving (17.8) in Case (ii).

In Case (iii), the elements ϕ and f from the von Neumann decomposition (17.7) are the two-component vector functions

$$\phi = \begin{pmatrix} \phi_\infty \\ \phi_\ell \end{pmatrix} \quad \text{and} \quad f = \begin{pmatrix} f_\infty \\ f_\ell \end{pmatrix}.$$

From (17.7) it follows that

$$\phi_\infty(x) = \alpha\xi\sqrt{1-k^2}e^x\chi_{(-\infty,0)}(x)$$
$$- \beta\xi\sqrt{1-k^2}e^{\ell-x}\chi_{(0,\infty)}(x) + f_\infty(x), \quad x \in \mathbb{R},$$

and

$$\phi_\ell(x) = \alpha\xi e^x + \beta\xi k e^{\ell-x} + f_\ell(x), \quad x \in [0,\ell),$$

where the norming constant ξ is given by

$$\xi = \sqrt{\frac{2}{e^{2\ell} - k^2}}.$$

In particular,

$$\phi_\infty(0-) = \alpha\xi\sqrt{1-k^2} + f_\infty(0-),$$
$$\phi_\infty(0+) = -\beta\xi\sqrt{1-k^2}e^\ell + f_\infty(0+),$$
$$\phi_\ell(\ell) = \alpha\xi e^\ell + k\beta\xi + f_\ell(\ell).$$

Since $f \in \mathrm{Dom}(\dot{D})$, the boundary conditions

$$f_\infty(0+) = kf_\infty(0-),$$
$$f_\ell(0) = \sqrt{1-k^2}f_\infty(0-),$$
$$f_\ell(\ell) = 0$$

hold and hence

$$\phi_\infty(0-) = \alpha\xi\sqrt{1-k^2} + \gamma,$$
$$\phi_\infty(0+) = -\beta\xi\sqrt{1-k^2}e^\ell + k\gamma,$$
$$\phi_\ell(\ell) = \alpha\xi e^\ell + k\beta\xi,$$

where we use the shorthand notation

$$\gamma = f_\infty(0-).$$

    Combining the obtained equations we arrive at the following system of equations

$$\begin{pmatrix} \sqrt{1-k^2} & 0 & 1 \\ 0 & -e^\ell\sqrt{1-k^2} & k \\ e^\ell & k & 0 \end{pmatrix} \begin{pmatrix} \alpha\xi \\ \beta\xi \\ \gamma \end{pmatrix} = \begin{pmatrix} x \\ y \\ z \end{pmatrix},$$

where

$$\begin{pmatrix} x \\ y \\ z \end{pmatrix} = \begin{pmatrix} \phi_\infty(0-) \\ \phi_\infty(0+) \\ \phi_\ell(\ell) \end{pmatrix}.$$

Taking into account that the inverse matrix of the system is of the form

$$\frac{1}{(e^{2\ell} - k^2)\sqrt{1-k^2}} \begin{pmatrix} -k^2 & k & e^\ell\sqrt{1-k^2} \\ ke^\ell & -e^\ell & -k\sqrt{1-k^2} \\ e^{2\ell}\sqrt{1-k^2} & -k\sqrt{1-k^2} & -e^\ell(1-k^2) \end{pmatrix}$$

one easily obtains that

$$\alpha\xi = \frac{-k^2 x + ky + e^\ell\sqrt{1-k^2}\,z}{(e^{2\ell} - k^2)\sqrt{1-k^2}} \quad \text{and} \quad \beta\xi = \frac{ke^\ell x - e^\ell y - k\sqrt{1-k^2}\,z}{(e^{2\ell} - k^2)\sqrt{1-k^2}}.$$

Therefore,

$$\begin{aligned}
\alpha + \beta &= \xi^{-1} \frac{(ke^\ell - k^2)x + (k - e^\ell)y + (e^\ell - k)\sqrt{1-k^2}\,z}{(e^{2\ell} - k^2)\sqrt{1-k^2}} \\
&= \xi^{-1} \frac{(ke^\ell - k^2)\phi_\infty(0-) + (k - e^\ell)\phi_\infty(0+) + (e^\ell - k)\sqrt{1-k^2}\,\phi_\ell(\ell)}{(e^{2\ell} - k^2)\sqrt{1-k^2}} \\
&= \xi^{-1} \frac{1}{(e^\ell + k)} \left(\phi_\ell(\ell) - \frac{\phi_\infty(0+) - k\phi_\infty(0-)}{\sqrt{1-k^2}} \right) \\
&= \frac{1}{\sqrt{2}} \sqrt{\tanh\frac{\ell + \ell'}{2}} \left(\phi_\ell(\ell) - \frac{\phi_\infty(0+) - k\phi_\infty(0-)}{\sqrt{1-k^2}} \right),
\end{aligned}$$

which completes the proof of (17.8) in Case (iii). $\square$

The main result of this chapter is the following

Theorem 17.4 (EXPONENTIAL DECAY-QUANTUM ZENO EFFECT ALTERNATIVE). *Suppose that a metric graph* $\mathbb{Y}$ *is in one of the Cases (i)–(iii). Let* $\dot{D}$ *be the symmetric differentiation operator given by (4.1), (4.2) and (4.3), respectively. Assume, in addition, that* $\phi \in \mathrm{Dom}((\dot{D})^*)$, $\|\phi\| = 1$.
Let $H = D_\Theta$, $|\Theta| = 1$, *be the (magnetic) Hamiltonian referred to in Theorem 5.1.*

Then

$$\lim_{n \to \infty} |(e^{it/nH}\phi, \phi)|^{2n} = e^{-\tau(\Theta)|t|}, \quad t \in \mathbb{R}, \qquad (17.9)$$

where the decay constant $\tau(\Theta)$ is given by

$$\tau(\Theta) = \begin{cases} |\Theta\phi_\infty(0-) + \phi_\infty(0+)|^2, & \text{in Case } (i) \\ |\Theta\phi_\ell(\ell) + \phi_\ell(0)|^2, & \text{in Case } (ii) \\ \left|\Theta\phi_\ell(\ell) - \dfrac{\phi_\infty(0+) - k\phi_\infty(0-)}{\sqrt{1-k^2}}\right|^2, & \text{in Case } (iii). \end{cases} \qquad (17.10)$$

Here $0 < k < 1$ is the quantum gate coefficient from the boundary condition (4.3) that determines the symmetric operator $\dot{D}$ in Case (iii).

In particular, the state $\phi \in \mathcal{L} = \mathrm{Dom}((\dot{D})^*)$ is a resonant state under continuous monitoring of the quantum unitary evolution $\phi \to e^{itH}\phi$ if and only if

$$\phi \notin \mathrm{Dom}(H) = \mathrm{Dom}(D_\Theta).$$

Otherwise, the state ϕ is a Zeno state.

Proof. *Part 1.* First, we prove the assertion in the particular case of $\Theta = 1$, where the Hamiltonian H is given by the differentiation operator D, i.e.,

$$H = D = D_\Theta|_{\Theta=1}.$$

Since $\phi \in \mathrm{Dom}((\dot{D})^*)$, by the von Neumann formula, the element ϕ admits a unique decomposition

$$\phi = \alpha g_+ + \beta g_- + f, \qquad (17.11)$$

where $\alpha, \beta \in \mathbb{C}$, $f \in \mathrm{Dom}(\dot{D})$. Here we choose the deficiency elements $g_\pm$ to be given by (4.6), (4.7), and finally by (4.8) and (4.9) whenever the graph $\mathbb{Y}$ is in Cases (i), (ii), and (iii), respectively.

Without loss of generality, we may assume that the operator $H = D$ is already realized in its model representation in the Hilbert space $L^2(\mathbb{R}; d\mu)$ as the operator of multiplication by independent variable with the measure $\mu(d\lambda)$ determined by (17.5).

Indeed, the Weyl-Titchmarsh function $M_{(\dot{D},D)}(z)$ associated with the pair $(\dot{D}, D)$ and given by (7.1) admits the representation (17.4) with the measure $\mu(d\lambda)$ from (17.5). By Lemma 4.4, the symmetric differentiation operator $\dot{D}$ is prime and therefore the Hamiltonian $H = D$ is unitarily equivalent to its model representation in the Hilbert space $L^2(\mathbb{R}; d\mu)$ by Theorem C.1 in Appendix C. By Lemma 6.1,

$$g_+ - g_- \in \text{Dom}(H) = \text{Dom}(D).$$

Therefore, one can also assume that the decomposition (17.11) takes place in the model Hilbert space $L^2(\mathbb{R}; d\mu)$, where the deficiency elements $g_\pm$ are given by the partial fractions (see Remark C.2 in Appendix C)

$$g_\pm = \frac{1}{\lambda \mp i}, \quad \lambda \in \mathbb{R} \ \mu - \text{a.e.,}$$

and

$$f \in L^2(\mathbb{R}; (1 + \lambda^2)d\mu(\lambda)). \tag{17.12}$$

The spectral measure $(\mathsf{E}_H(d\lambda)\phi, \phi)$ of the element ϕ can be evaluated as follows

$$(\mathsf{E}_H(\delta)\phi, \phi) = \int_\delta \left| \alpha \frac{1}{\lambda - i} + \beta \frac{1}{\lambda + i} + f(\lambda) \right|^2 d\mu(\lambda),$$

with $\delta \subset \mathbb{R}$ a Borel set.
Therefore,

$$\begin{aligned}
(\mathsf{E}_H(\delta)\phi, \phi) = {} & |\alpha + \beta|^2 \int_\delta \frac{d\mu(\lambda)}{\lambda^2 + 1} \\
& + 2\text{Re}\,\alpha\overline{\beta} \int_\delta \left(\frac{1}{(\lambda - i)^2} - \frac{1}{\lambda^2 + 1} \right) d\mu(\lambda) \\
& + 2\text{Re} \int_\delta \left(\alpha \frac{1}{\lambda - i} + \beta \frac{1}{\lambda + i} \right) \overline{f(\lambda)} d\mu(\lambda) \\
& + \int_\delta |f(\lambda)|^2 \, d\mu(\lambda).
\end{aligned} \tag{17.13}$$

It turns out that the first term in (17.13) determines the leading term of the asymptotics in the heavy-tailed distribution of the spectral measure $(\mathsf{E}_H(d\lambda)\phi, \phi)$ whenever

$$\alpha + \beta \neq 0.$$

Indeed, we have

$$I =: \int_{|s|>\lambda} \left| \frac{1}{(s-i)^2} - \frac{1}{s^2+1} \right| d\mu(s) \leq \frac{2}{\lambda} \int_{|s|>\lambda} \frac{d\mu(s)}{s^2+1}.$$

By Lemma 17.1, the following limit exists,

$$\lim_{\lambda \to +\infty} \lambda \int_{|s|>\lambda} \frac{d\mu(s)}{s^2+1} < \infty$$

and therefore

$$I = O(\lambda^{-2}) \quad \text{as } \lambda \to +\infty.$$

Next,

$$II =: \left| \int_{|s|>\lambda} \frac{\overline{f(s)}}{s \pm i} d\mu(s) \right|$$

$$\leq \frac{1}{\lambda} \sqrt{\int_{|s|>\lambda} \frac{d\mu(s)}{s^2+1}} \cdot \sqrt{\int_{|s|>\lambda}^{\infty} (1+s^2)|f(s)|^2 d\mu(s)}$$

$$= o(\lambda^{-3/2}) \quad \text{as } \lambda \to +\infty,$$

where we have used (17.12).

Finally,

$$III =: \int_{|s|>\lambda} |f(s)|^2 \, d\mu(s)$$

$$\leq \frac{1}{\lambda^2} \int_{|s|>\lambda} (1+s^2)|f(s)|^2 d\mu(s) = o(\lambda^{-2}) \quad \text{as } \lambda \to \infty.$$

Therefore,

$$I + II + III = o(\lambda^{-3/2}) \quad \text{as } \lambda \to +\infty$$

and from (17.13) we obtain

$$\lambda(\mathsf{E}_H((\lambda,\infty))\phi,\phi) = |\alpha+\beta|^2 \lambda \int_{\lambda}^{\infty} \frac{d\mu(s)}{s^2+1} + o(\lambda^{-1/2}) \quad \text{as } \lambda \to \infty.$$

In a similar way one proves that

$$\lambda(\mathsf{E}_H((-\infty,-\lambda))\phi,\phi) = |\alpha+\beta|^2 \lambda \int_{-\infty}^{-\lambda} \frac{d\mu(s)}{s^2+1} + o(\lambda^{-1/2}) \quad \text{as } \lambda \to \infty.$$

By Lemma 17.1,

$$\lim_{\lambda \to +\infty} \lambda \int_\lambda^\infty \frac{d\mu(s)}{s^2 + 1}$$

$$= \lim_{\lambda \to +\infty} \lambda \int_{-\infty}^{-\lambda} \frac{d\mu(s)}{s^2 + 1} = \begin{cases} 1, & \text{in Case (i)} \\ \coth \dfrac{\ell}{2}, & \text{in Case (ii)} \\ \coth \dfrac{\ell + \ell'}{2}, & \text{in Case (iii).} \end{cases}$$

Therefore,

$$\lim_{\lambda \to +\infty} \lambda (\mathsf{E}_H((\lambda, \infty))\phi, \phi)$$

$$= \lim_{\lambda \to +\infty} \lambda (\mathsf{E}_H((-\infty, -\lambda))\phi, \phi)$$

$$= \frac{1}{\pi} |\alpha + \beta|^2 \begin{cases} 1, & \text{in Case (i)} \\ \coth \dfrac{\ell}{2}, & \text{in Case (ii)} \\ \coth \dfrac{\ell + \ell'}{2}, & \text{in Case (iii).} \end{cases}$$

On the other hand, from Lemma 17.3 it follows that

$$|\alpha + \beta|^2 = \frac{1}{2} \begin{cases} |\phi_\infty(0-) + \phi_\infty(0+)|^2, & \text{in Case (i)} \\ \tanh \dfrac{\ell}{2} |\phi_\ell(0) + \phi_\ell(\ell)|^2, & \text{in Case (ii)} \\ \tanh \dfrac{\ell + \ell'}{2} \left| \phi_\ell(\ell) - \dfrac{\phi_\infty(0+) - k\psi_\infty(0-)}{\sqrt{1 - k^2}} \right|^2, & \text{in Case (iii).} \end{cases}$$

$$(17.14)$$

Here, in Case (iii),

$$\ell' = \log \frac{1}{k}.$$

Hence

$$\lim_{\lambda \to +\infty} \lambda (\mathsf{E}_H((\lambda, \infty))\phi, \phi)$$

$$= \lim_{\lambda \to +\infty} \lambda (\mathsf{E}_H((-\infty, -\lambda))\phi, \phi)$$

$$= \frac{1}{2\pi} \begin{cases} |\phi_\infty(0-) + \phi_\infty(0+)|^2, & \text{in Case (i)} \\ |\phi_\ell(0) + \phi_\ell(\ell)|^2, & \text{in Case (ii)} \\ \left| \phi_\ell(\ell) - \dfrac{\phi_\infty(0+) - k\phi_\infty(0-)}{\sqrt{1 - k^2}} \right|^2, & \text{in Case (iii).} \end{cases}$$

To complete the proof of (17.10) in case $\Theta = 1$ it remains to apply Theorem 15.12.

Part 2. Now we can treat the general case of an arbitrary Θ, $|\Theta| = 1$.

Let $H = D_\Theta$ be the magnetic Hamiltonian in Case (i). Denote by U_Θ the unitary operator in $L^2(\mathbb{Y}) = L^2(\mathbb{R})$ defined as

$$(U_\Theta f)(x) = \overline{\Theta}\chi_{(-\infty,0)}(x)f(x) + \chi_{(0,\infty)}(x)f(x).$$

One verifies that

$$D_\Theta = U_\Theta D U_\Theta^*$$

and therefore

$$\lim_{n\to\infty} |(e^{it/nH}\phi, \phi)|^{2n} = \lim_{n\to\infty} |(e^{it/nD_\Theta}\phi, \phi)|^{2n} = \lim_{n\to\infty} |(e^{it/nD}U_\Theta^*\phi, U_\Theta^*\phi)|^{2n}.$$

By Part 1 of the proof,

$$\lim_{n\to\infty} |(e^{it/nD}U_\Theta^*\phi, U_\Theta^*\phi)|^{2n} = \exp\left(-|(U_\Theta^*\phi_\infty)(0+) + (U_\Theta^*\phi_\infty)(0-)|^2|t|\right)$$

$$= \exp\left(-|\phi_\infty(0+) + \Theta\phi_\infty(0-)|^2|t|\right).$$

Therefore,

$$\tau(\Theta) = |\phi_\infty(0+) + \Theta\phi_\infty(0-)|^2,$$

which proves (17.10) in Case (i).

In Cases (ii) and (iii), we have the commutation relation (cf. (5.6))

$$D_\Theta = U_\Theta \left(D + \frac{\arg\Theta}{\ell}I\right) U_\Theta^*,$$

where U_Θ the unitary multiplication operator in $L^2(\mathbb{Y})$ given by

$$(U_\Theta\phi(x)) = e^{-i\frac{\arg\Theta}{\ell}x}\phi(x), \quad x \in \mathbb{Y}.$$

Therefore,

$$\lim_{n\to\infty} |(e^{it/nH}\phi, \phi)|^{2n} = \lim_{n\to\infty} |e^{it\frac{\arg\Theta}{n\ell}}(e^{it/nD}U^*\phi, U^*\phi)|^{2n}$$

$$= \lim_{n\to\infty} |(e^{it/nD}U^*\phi, U^*\phi)|^{2n}.$$

Now the claim (17.10) follows by applying the result of Part 1 of the proof to the state $U^*\phi$.

To complete the proof it remains to show that under the hypothesis that $\phi \in \mathrm{Dom}((\dot{D})^*)$ the decay constant $\tau(\Theta)$ vanishes if and only if $\phi \in \mathrm{Dom}(D_\Theta)$. Indeed, if $\phi \in \mathrm{Dom}(D_\Theta) \subset \mathrm{Dom}(|D_\Theta|^{1/2})$, then ϕ is a Zeno state (by Proposition 15.3) and hence $\tau(\Theta) = 0$. One can also see right away that the boundary conditions (5.1), (5.2) and (5.3) imply $\tau(\Theta) = 0$.

The converse (under the hypothesis that $\phi \in \mathrm{Dom}(\dot{D})^*)$) is also true. It is obvious in Cases (i) and (ii). In Case (iii) the equality $\tau(\Theta) = 0$ implies

$$\Theta \phi_\ell(\ell) = \frac{\phi_\infty(0+) - k\phi_\infty(0-)}{\sqrt{1-k^2}} \tag{17.15}$$

and since $\phi \in \mathrm{Dom}((\dot{D})^*)$, from (4.4) it also follows that

$$\phi_\infty(0-) - k\,\phi_\infty(0+) - \sqrt{1-k^2}\,\phi_\ell(0) = 0. \tag{17.16}$$

Multiplying (17.15) by k and using (17.16) we obtain

$$
\begin{aligned}
k\Theta\phi_\ell(\ell) &= \frac{k\phi_\infty(0+) - k^2\phi_\infty(0-)}{\sqrt{1-k^2}} \\
&= \frac{\phi_\infty(0-) - \sqrt{1-k^2}\,\phi_\ell(0+) - k^2\phi_\infty(0-)}{\sqrt{1-k^2}} \\
&= \sqrt{1-k^2}\phi_\infty(0-) - \phi_\ell(0),
\end{aligned}
$$

which shows that

$$\phi_\ell(0) = \sqrt{1-k^2}\phi_\infty(0-) - k\Theta\phi_\ell(\ell). \tag{17.17}$$

From (17.15) and (17.17) we get

$$
\begin{pmatrix} \phi_\infty(0+) \\ \phi_\ell(0) \end{pmatrix} = \begin{pmatrix} k & \sqrt{1-k^2}\Theta \\ \sqrt{1-k^2} & -k\Theta \end{pmatrix} \begin{pmatrix} \phi_\infty(0-) \\ \phi_\ell(\ell) \end{pmatrix}
$$

and hence (5.3) holds proving that $\phi \in \mathrm{Dom}(D_\Theta)$.   □

Let $\widehat{D}$ be the maximal dissipative differential operator defined by (8.1) (with $k = 0$) whenever the graph $\mathbb{Y}$ is in Case (i) and by (8.4) and (8.5) whenever the graph $\mathbb{Y}$ is in Cases (ii) and (iii), respectively. Assume, in addition, that the initial state ϕ is such that $\phi \in \mathrm{Dom}(\widehat{D}) \cup \mathrm{Dom}(\widehat{D}^*)$.

The following lemma shows that under these assumptions the decay rate of the state ϕ under continuous monitoring of the unitary evolution

$$\phi \mapsto e^{itH}\phi \tag{17.18}$$

is determined by the state only and is, in fact, independent of the self-adjoint realization $H = D_\Theta$ of the symmetric differentiation operator $\dot{D}$.

Lemma 17.5. *Assume the hypothesis of Theorem 17.4. Let $\widehat{D}$ be the maximal dissipative differential operator defined by (8.1) (with $k = 0$) whenever the graph $\mathbb{Y}$ is in Case (i) and by (8.4) and (8.5) whenever the graph $\mathbb{Y}$ is in Cases (ii) and (iii), respectively.*

Then the decay constant $\tau = \tau(\Theta)$ given by (17.10) does not depend on Θ if and only if

$$\phi \in \mathrm{Dom}(\widehat{D}) \cup \mathrm{Dom}((\widehat{D})^*).$$

In this case,

$$\tau = \begin{cases} |\phi_\infty(0-)|^2, & \text{in Case (i)} \\ |\phi_\ell(\ell)|^2, & \text{in Case (ii)} \quad \text{whenever } \phi \in \mathrm{Dom}(\widehat{D}), \quad (17.19) \\ |\phi_\ell(\ell)|^2, & \text{in Case (iii),} \end{cases}$$

and

$$\tau = \begin{cases} |\phi_\infty(0+)|^2, & \text{in Case (i)} \\ |\phi_\ell(0)|^2, & \text{in Case (ii)} \quad \text{whenever } \phi \in \mathrm{Dom}((\widehat{D})^*). \\ \dfrac{|\phi_\infty(0+) - k\phi_\infty(0-)|^2}{1 - k^2}, & \text{in Case (iii),} \end{cases}$$

$$(17.20)$$

Here, in Case (iii), k is the quantum gate coefficient.

Proof. It is easily seen from (17.10) that the decay constant $\tau(\Theta)$ does not depend on Θ if and only if either

$$\begin{cases} \phi_\infty(0+) = 0, & \text{in Case (i)} \\ \phi_\ell(0) = 0, & \text{in Case (ii)} \quad (17.21) \\ \phi_\infty(0+) = k\phi_\infty(0-), & \text{in Case (iii),} \end{cases}$$

or

$$\begin{cases} \phi_\infty(0-) = 0, & \text{in Case (i)} \\ \phi_\ell(\ell) = 0, & \text{in Case (ii)} \quad (17.22) \\ \phi_\ell(\ell) = 0, & \text{in Case (iii),} \end{cases}$$

or both.

Recall that the boundary conditions (8.1) (with $k = 0$), (8.4), and (8.5) for the dissipative differentiation operator $\widehat{D}$ yield

$$\phi_\infty(0+) = 0 \quad \text{(in Case (i))}$$
$$\phi_\ell(0) = 0 \quad \text{(in Case (ii))} \tag{17.23}$$

and

$$\begin{cases} \phi_\infty(0+) = k\phi_\infty(0-) \\ \phi_\ell(0) = \sqrt{1 - k^2}\phi_\infty(0-) \end{cases} \quad \text{(in Case (iii))}. \tag{17.24}$$

Notice that in Case (iii) the first condition in (17.24) implies the second one whenever $\phi \in \text{Dom}((\dot{D})^*)$. Indeed, the membership $\phi \in \text{Dom}((\dot{D})^*)$ means that

$$\phi_\infty(0-) - k\phi_\infty(0+) - \sqrt{1 - k^2}\phi_\ell(0) = 0$$

by Lemma 4.2 and the claim follows by a simple computation.

Now, it is straightforward to see that under the hypothesis that $\phi \in \mathcal{L} = \text{Dom}((\dot{D})^*)$ the boundary conditions (17.21) hold if and only if $\phi \in \text{Dom}(\widehat{D})$. In this case (17.19) follows from (17.10) in Theorem 17.4.

Next, the boundary conditions for the adjoint operator $(\widehat{D})^*$ (9.13) (with $k = 0$), (9.14), and (9.15) can be rewritten as

$$\phi_\infty(0-) = 0 \quad \text{(in Case (i))}$$
$$\phi_\ell(\ell) = 0 \quad \text{(in Case (ii))} \tag{17.25}$$

and

$$\begin{cases} \phi_\infty(0-) &= k\phi_\infty(0+) + \sqrt{1 - k^2}\phi_\ell(0) \\ \phi_\ell(\ell) &= 0 \end{cases} \quad \text{(in Case (iii))}. \tag{17.26}$$

By Lemma 4.2, the first condition in (17.26) simply means that $\phi \in \text{Dom}((\dot{D})^*)$. Therefore, the boundary conditions (17.22) hold if and only if and $\phi \in \text{Dom}(\widehat{D}^*)$, In this case (17.20) follows from (17.10) in Theorem 17.4. $\square$

If the initial state ϕ is taken from the somewhat narrower test space

$$\mathcal{M} = \text{Dom}(\widehat{D}) \cup \text{Dom}((\widehat{D}))^* \subset \text{Dom}((\dot{D})^*) = \mathcal{L},$$

then Lemma 17.5 states that continuous monitoring of the unitary evolution (17.18) is universal (in the sense that the corresponding decay rate

τ referred to in Lemma 17.5 is independent of the choice of the magnetic Hamiltonian H).

In this case, i.e. if $\phi \in \mathcal{M}$, the universal exponent τ can also be recognized as the decay rate associated with continuous monitoring of the unitary evolution

$$\widehat{\phi} \mapsto e^{it\mathbb{H}}\widehat{\phi} \tag{17.27}$$

in an extended Hilbert $\mathfrak{H}$ containing $L^2(\mathbb{Y})$ as a proper subspace. Here $\mathfrak{H} = L^2(\mathbb{X})$, where $\mathbb{X}$ is the full metric graph: $\mathbb{X} = \mathbb{Y} \sqcup \mathbb{Y}$ if the metric graph $\mathbb{Y}$ in Cases (i) and (iii), and $\mathbb{X}$ can be identified with $\mathbb{R}$ if $\mathbb{Y} = (0, \ell)$ is in Case (ii), the Hamiltonian $\mathbb{H}$ is a self-adjoint dilation of the dissipative differentiation operator $\widehat{D}$, and the new state $\widehat{\phi} \in L^2(\mathbb{X})$ of the extended quantum system is identified with the initial state ϕ being naturally imbedded to the space $\mathfrak{H} = L^2(\mathbb{X})$.

The precise statement is as follows.

Corollary 17.6. *Let $\mathbb{H}$ be a self-adjoint dilation in the Hilbert space $\mathfrak{H} = L^2(\mathbb{X})$ of the dissipative differentiation operator $\widehat{D}$. Assume that*

$$\phi \in \mathcal{M} = \mathrm{Dom}(\widehat{D}) \cup \mathrm{Dom}(\widehat{D}^*), \quad \|\phi\| = 1.$$

Denote by $\widehat{\phi}$ a state in $\mathfrak{H}$ such that

$$\phi = P_{L^2(\mathbb{Y})}\widehat{\phi} \quad and \quad (I - P_{L^2(\mathbb{Y})})\widehat{\phi} = 0,$$

where $P_{L^2(\mathbb{Y})}$ is the orthogonal projection from the Hilbert space $L^2(\mathbb{X})$ onto its subspace $L^2(\mathbb{Y})$.

Then

$$\lim_{n\to\infty} |(e^{it/n\mathbb{H}}\widehat{\phi}, \widehat{\phi})|^{2n} = e^{-\tau|t|}, \quad t \in \mathbb{R}, \tag{17.28}$$

where the decay constant τ is given by (17.19) if $\phi \in \mathrm{Dom}(\widehat{D})$ and by (17.20) if $\phi \in \mathrm{Dom}((\widehat{D})^*)$, respectively.

Proof. Suppose first that $\phi \in \mathrm{Dom}(\widehat{D})$.

Denote by $\mathcal{D}$ the differentiation operator $i\frac{d}{dx}$ on

$$\mathrm{Dom}(\mathcal{D}) = \bigoplus_{e \subset \mathbb{Y}} W_2^1(e),$$

where the sum is taken over all edges e of the graph $\mathbb{Y}$.

Integration by parts for $\phi \in \text{Dom}(\mathcal{D})$ yields

$$\text{Im}(\mathcal{D}\phi, \phi) = \frac{1}{2} \begin{cases} |\phi_\infty(0-)|^2 - |\phi_\infty(0+)|^2 \\ |\phi_\ell(\ell)|^2 - |\phi_\ell(0)|^2 \\ |\phi_\infty(0-)|^2 - |\phi_\infty(0+)|^2 + |\phi_\ell(\ell)|^2 - |\phi_\ell(0)|^2 \end{cases} \tag{17.29}$$

in Cases (i), (i) and (iii), respectively. Since $\phi \in \text{Dom}(\widehat{D})$, taking into account the boundary conditions (17.23) and (17.24), from (17.29) we obtain

$$\text{Im}(\widehat{D}\phi, \phi) = \text{Im}(\mathcal{D}\phi, \phi) = \frac{1}{2} \begin{cases} |\phi_\infty(0-)|^2, & \text{in Case (i)} \\ |\phi_\ell(\ell)|^2, & \text{in Case (ii)} \\ |\phi_\ell(\ell)|^2, & \text{in Case (iii).} \end{cases}$$

Therefore,

$$2\text{Im}(\widehat{D}\phi, \phi) = \tau,$$

where τ is given by (17.19).

By Lemma 15.15,

$$\lim_{n \to \infty} |(e^{it\widehat{D}}\phi, \phi)|^{2n} = e^{-2\text{Im}(\widehat{D}\phi, \phi)t}, \quad t \geq 0, \quad \phi \in \text{Dom}(\widehat{D}),$$

and therefore

$$\lim_{n \to \infty} |(e^{it/n\widehat{D}}\phi, \phi)|^{2n} = e^{-\tau t}, \quad t \geq 0, \quad \phi \in \text{Dom}(\widehat{D}), \tag{17.30}$$

where τ given by (17.19).

Since $\mathbb{H}$ dilates $\widehat{D}$, we have

$$(e^{it/n\mathbb{H}}\widehat{\phi}, \widehat{\phi}) = (e^{it\widehat{D}}\phi, \phi), \quad t \geq 0, \tag{17.31}$$

and therefore

$$\lim_{n \to \infty} |(e^{it/n\mathbb{H}}\widehat{\phi}, \widehat{\phi})|^{2n} = \lim_{n \to \infty} |(e^{it/n\widehat{D}}\phi, \phi)|^{2n} = e^{-\tau t}, \quad t \geq 0. \tag{17.32}$$

To complete the proof of (17.28) for $\phi \in \text{Dom}(\widehat{D})$ it remains to observe that the return probability $p(t) = |(e^{it/n\mathbb{H}}\widehat{\phi}, \widehat{\phi})|^2$ is an even function in t.

Next, suppose that $\phi \in \text{Dom}((\widehat{D})^*)$.

As above, by Lemma 15.15,

$$\lim_{n\to\infty} |(e^{-it/n(\widehat{D})^*}\phi,\phi)|^{2n} = e^{2\mathrm{Im}((\widehat{D})^*\phi,\phi)t}, \quad t \geq 0. \tag{17.33}$$

Since $\phi \in \mathrm{Dom}((\widehat{D})^*)$, one can use boundary conditions (17.25), (17.26) and the equality (17.29) to obtain that

$$\mathrm{Im}((-\widehat{D})^*\phi,\phi) = \mathrm{Im}((-\mathcal{D})\phi,\phi)$$

$$= \frac{1}{2} \begin{cases} |\phi_\infty(0+)|^2, & \text{in Case (i)} \\ |\phi_\ell(0)|^2, & \text{in Case (ii)} \\ |\phi_\infty(0-)|^2 - |\phi_\infty(0+)|^2 - |\phi_\ell(0)|^2, & \text{in Case (iii)} . \end{cases}$$

(In Case (iii), we took into account the second condition in (17.26) that $\phi_\ell(\ell) = 0$).

Now, using the first condition in (17.26), one computes

$$|\phi_\infty(0-)|^2 - |\phi_\infty(0+)|^2 - |\phi_\ell(0)|^2 = \left| \frac{\phi_\infty(0+) - k\phi_\infty(0-)}{\sqrt{1-k^2}} \right|^2,$$

which shows that

$$2\mathrm{Im}((\widehat{D})^*\phi,\phi) = - \begin{cases} |\phi_\infty(0+)|^2, & \text{in Case (i)} \\ |\phi_\ell(0)|^2, & \text{in Case (ii)} \\ \left| \dfrac{\phi_\infty(0+) - k\phi_\infty(0-)}{\sqrt{1-k^2}} \right|^2, & \text{in Case (iii)} \end{cases} = \tau,$$

where τ is given by (17.20).

Again, by Lemma 15.15,

$$\lim_{n\to\infty} |(e^{-it(\widehat{D})^*}\phi,\phi)|^{2n} = e^{2\mathrm{Im}((\widehat{D})^*\phi,\phi)t}, \quad t \geq 0, \quad \phi \in \mathrm{Dom}((\widehat{D})^*),$$

and therefore

$$\lim_{n\to\infty} |(e^{-it/n(\widehat{D})^*}\phi,\phi)|^{2n} = e^{-\tau t}, \quad t \geq 0, \quad \phi \in \mathrm{Dom}((\widehat{D})^*), \tag{17.34}$$

where τ given by (17.20).

Since $\mathbb{H}$ dilates $\widehat{D}$, we obtain

$$\lim_{n\to\infty} |(e^{-it/n(\widehat{D})^*}\phi, \phi)|^{2n}$$

$$= \lim_{n\to\infty} |(\phi, (e^{-it/n(\widehat{D})^*})^*\phi)|^{2n} = \lim_{n\to\infty} |(e^{it/n\widehat{D}}\phi, \phi)|^{2n}$$

$$= \lim_{n\to\infty} |(e^{it/n\mathbb{H}}\phi, \phi)|^{2n} = e^{-\tau t}, \quad t > 0,$$

which proves (17.28) for $\phi \in \mathrm{Dom}((\widehat{D})^*)$ with τ given by (17.20) and then, by symmetry, for all $t \in \mathbb{R}$. $\qquad\square$

Remark 17.7. We remark that if $\phi \in \mathrm{Dom}(\widehat{D}) \cap \mathrm{Dom}((\widehat{D})^*) = \mathrm{Dom}(\dot{D})$ and therefore $\phi \in \mathrm{Dom}(D_\Theta)$ for all $|\Theta| = 1$, then ϕ is a Zeno state under continuous monitoring of the dynamics $\phi \mapsto e^{itH}\phi$ for any self-adjoint realizations $H = D_\Theta$ of the differentiation operator. Therefore, in this case the corresponding decay constant $\tau = \tau(\Theta) = 0$ is Θ-independent for an obvious reason.

Also notice that under the requirement that $\phi \in \mathcal{M} = \mathrm{Dom}(\widehat{D}) \cup \mathrm{Dom}((\widehat{D})^*)$ the boundary data that determine the decay constant (17.19) and (17.20) can also be evaluated as follows.

Assume, for instance, that the metric graph $\mathbb{Y}$ is in Case (iii) with its main vertex at the origin ($\mu = 0$). Denote by $\mathbb{X}$ the full metric graph containing $\mathbb{Y}$ as its subgraph and let $\mathbb{H}$ be the self-adjoint dilation in the extended Hilbert space $L^2(\mathbb{X})$ of the dissipative operator $\widehat{D}$ in $L^2(\mathbb{Y})$.

Suppose that a two-component vector-function $\Psi = \begin{pmatrix} \phi_\uparrow \\ \phi_\downarrow \end{pmatrix} \in L^2(\mathbb{X})$ is a continuation of the function ϕ from the graph $\mathbb{Y}$ onto the full graph $\mathbb{X}$,

$$\Psi(x) = \phi(x), \quad x \in \mathbb{Y}, \tag{17.35}$$

such that $\Psi \in \mathrm{Dom}(\mathbb{H})$.

Then the decay constant (17.19) can be evaluated via the second component of the vector-function Ψ as

$$\tau = |\phi_\downarrow(\ell)|^2.$$

Indeed, since $\Psi \in \mathrm{Dom}(\mathbb{H})$, the two-component vector-function $\Psi(x)$ is continuous, so is its second component $\phi_\downarrow(x)$. In particular, $\Psi(\ell) = \phi(\ell)$, so that

$$\phi_\ell(\ell) = \phi_\downarrow(\ell) \tag{17.36}$$

and hence

$$\tau = |\phi_\ell(\ell)|^2 = |\phi_\downarrow(\ell)|^2.$$

Moreover, for the decay constant (17.20) we have a similar expression

$$\tau = |\phi_\downarrow(0-)|^2. \tag{17.37}$$

Indeed, since $\Psi \in \mathrm{Dom}(\mathbb{H})$, by (14.5) we get

$$\begin{pmatrix} \phi_\uparrow(0+) \\ \phi_\downarrow(0+) \end{pmatrix} = \begin{pmatrix} k & -\sqrt{1-k^2} \\ \sqrt{1-k^2} & k \end{pmatrix} \begin{pmatrix} \phi_\uparrow(0-) \\ \phi_\downarrow(0-) \end{pmatrix}.$$

In particular,

$$\phi_\uparrow(0+) = k\phi_\uparrow(0-) - \sqrt{1-k^2}\phi_\downarrow(0-).$$

Hence

$$\phi_\downarrow(0-) = \frac{k\phi_\uparrow(0-) - \phi_\uparrow(0+)}{\sqrt{1-k^2}} = \frac{k\phi_\infty(0-) - \phi_\infty(0+)}{\sqrt{1-k^2}}. \tag{17.38}$$

By (17.20),

$$\tau = \left| \frac{k\phi_\infty(0-) - \phi_\infty(0+)}{\sqrt{1-k^2}} \right|^2,$$

which together with (17.38) proves (17.37).

Chapter 18

PRELIMINARIES: PROBABILITIES VERSUS AMPLITUDES

To discuss applications of the continuous monitoring principle in connection with the exponential decay phenomenon in quantum mechanics, we need to warm up with some preliminaries.

Recall that "when we deal with probabilities under ordinary circumstances, there are the following "rules of composition": 1) if something can happen in alternative ways, we add the probabilities for each of the different ways: 2) if the event occurs as a succession of steps — or depends on a number of things happening —'concomitantly' (independently) — then we multiply the probabilities of each of the steps (or things)" [34, Ch. 3].

Apparently, under certain circumstances the probability P of an event that can be realized in two (at first glance mutually exclusive) alternative ways A_1 and A_2 is not necessarily equals the sum of probabilities P_1 and P_2 of the events A_1 and A_2, that is,

$$P \neq P_1 + P_2 \quad \text{(in general).}$$

A more detailed analysis of the experimental data shows that the concept of an alternative should be analyzed more carefully and one has to distinguish between *exclusive* and *interference alternatives*. The latter occurs if there is no (experimental) evidence available to answer the question of how the final event has been realized, via the occurrence of A_1 or A_2? In other words, "when alternatives cannot possibly be resolved by any experiment, they always interfere" [35, page 14].

If the alternative ways of a realization of the event are exclusive, one has the usual addition law of probabilities

$$P_{\mathrm{ex}} = P_1 + P_2. \tag{18.1}$$

In the case of an interference alternative, the rules of composition should be applied to the amplitudes of probability instead. Recall that there are complex numbers ϕ, ϕ_1, ϕ_2 (the probability amplitudes), obtained, for example, by solving a kind of wave equation, such that

$$P_{\mathrm{int}} = |\phi|^2, \quad P_1 = |\phi_1|^2 \quad \text{and} \quad P_2 = |\phi_2|^2.$$

In particular, the addition law for (probability) amplitudes

$$\phi = \phi_1 + \phi_2 \tag{18.2}$$

yields

$$P_{\mathrm{int}} = |\phi|^2 = |\phi_1 + \phi_2|^2 \quad (\neq P_{\mathrm{ex}} \text{ in general}). \tag{18.3}$$

In this context it should be stressed that the (experimental) knowledge of the probabilities $P_{1,2}$ (but not the amplitudes $\phi_{1,2}$) only gives the two sided-estimate for the probability P_{int} of the final event

$$P_1 + P_2 - 2\sqrt{P_1 P_2} \leq P_{\mathrm{int}} \leq \min\{1, P_1 + P_2 + 2\sqrt{P_1 P_2}\}.$$

All of this is well known and has been extensively discussed in detail in connection with the two slit experiment (see, e.g., [34, 35, 44], also see [28, 70] for the concept of interaction-free measurements).

Our goal is to provide a solid mathematical background for understanding the phenomenon on a simple one-dimensional example of a quantum system and develop a framework where the concepts of exclusive and interference alternatives can be rigorously discussed.

Chapter 19

MASSLESS PARTICLES
ON A RING

Consider a quantum system the configuration space of which is a ring $\mathcal{S}$ obtained by identifying the end-points of a finite interval $[0, \ell]$. The dynamics of the system is described by the strongly continuous group of unitary operators $U(t) = e^{-it/\hbar H}$, where the Hamiltonian H is given by the differentiation operator on the ring, or, equivalently, by the differentiation operator on the finite interval $[0, \ell]$ with periodic boundary conditions. That is,

$$H = ic\hbar\frac{d}{dx} \quad \text{on} \quad \text{Dom}(H) = \{f \in W_2^1((0, \ell)) \mid f(0) = f(\ell)\}. \quad (19.1)$$

To motivate the choice of the Hamiltonian we use the energy-momentum relation

$$E^2 = (cP)^2 + (mc^2)^2$$

and assume that we are dealing with a massless particle $(m = 0)$ and then choose a square root brunch of $(cP)^2$ to define the energy operator H as (cf. [130])

$$H = -cP.$$

Here P denotes the momentum of a particle moving with no dispersion at the speed of light on the ring $\mathcal{S}$ in the direction from "$x = 0$ to $x = \ell$."

In order to save ourselves from inventing new words such as "wavicles", we have chosen to call these objects "בּל-particles" (cf. [34, p. 85]). In our opinion, בּל-particles may, for instance, serve as a one-dimensional prototype of low energy electrons in the vicinity of an impurity in a zero-gap semiconductor. Recall that such electrons can formally be described by the two-dimensional Dirac-like Hamiltonian

$$H = -ic\hbar\,\boldsymbol{\sigma}\cdot\boldsymbol{\nabla} + V, \quad \text{with} \quad c = \nu_F,$$

where ν_F is the Fermi velocity, $\boldsymbol{\sigma} = (\sigma_x, \sigma_y)$ are the 2×2 conventional Pauli matrices and V is a short range "defect" potential [20]. In this simplified model we will imagine these electrons as fake spin-zero electrons which, however, can carry the charge $\mathfrak{e}$.

Suppose that $\phi \in L^2((0,\ell))$, $\|\phi\| = 1$, is a wave-function describing the initial state of the quantum system with the Hamiltonian

$$H = ic\hbar\frac{d}{dx}$$

with periodic boundary conditions (19.1). If no observation is made whatsoever, the time evolution

$$U(t)\phi = e^{-it/\hbar H}\phi$$

of the state ϕ is given by the family of unitary transformations

$$(U(t)\phi)(x) = \widetilde{\phi}(x + ct), \quad x \in [0,\ell], \quad t \in \mathbb{R}.$$

Here $\widetilde{\phi}$ denotes the periodic extension of the function $\phi(x)$ from the interval $[0,\ell]$ onto the full real axis. In other words, the wave packet $U(t)\phi$ is confined to move at the speed of light c without dispersion on the ring $\mathbb{S}$ of radius $\ell/2\pi$, obtained from the interval $[0,\ell]$ by identifying its end-points.

In this case, the survival probability

$$p(t) = |(e^{-it/\hbar H}\phi, \phi)|^2$$

to the initial state ϕ is a periodic function with the period $T = \ell/c$.

In the forthcoming chapters we will learn that under continuous monitoring the quantum system on the ring $\mathbb{S}$ becomes an open quantum system, the particles can be emitted and the whole system can be considered as a kind of quantum antenna.

Chapter 20

CONTINUOUS MONITORING WITH INTERFERENCE

Throughout this chapter we assume that the initial state ϕ is a $W_2^1((0,\ell))$-function that is allowed to have a discontinuity (jump) at the point of the observation $x = 0 \equiv \ell$, that is,

$$\phi(0) \neq \phi(\ell), \quad \text{in general.}$$

Notice that although $\phi \in W_2^1((0,\ell))$, the initial state ϕ is not required to belong to the domain of the Hamiltonian H in general. That is, it is not assumed that $\phi \in W_2^1(\mathbb{S})$, with $\mathbb{S}$ the ring obtained by identifying the end-points of the interval $[0, \ell]$.

The decay properties of states with a unique jump-point on the ring under continuous monitoring are described by the following result.

Theorem 20.1. *Suppose that* $\mathbb{Y} = (0, \ell)$ *is a metric graph in Case* (ii). *In the Hilbert space* $\mathcal{H} = L^2(\mathbb{Y})$ *denote by* H *the differentiation operator*

$$H = ic\hbar\frac{d}{dx} \quad on \quad \text{Dom}(H) = \{f \in W_2^1((0,\ell)) \,|\, f(0) = f(\ell)\}. \quad (20.1)$$

If $\phi \in W_2^1((0,\ell))$ *and* $\|\phi\| = 1$, *then*

$$\lim_{n \to \infty} |(e^{-it/(\hbar n)H}\phi, \phi)|^{2n} = e^{-\tau|t|}, \quad t \in \mathbb{R}, \quad (20.2)$$

where

$$\tau = c|\Delta\phi|^2 = c|\phi(\ell) - \phi(0)|^2. \quad (20.3)$$

Proof. Let D_Θ be the differentiation operator $i\frac{d}{dx}$ on the finite interval $[0, \ell]$ defined on

$$\mathrm{Dom}(D_\Theta) = \{f \in W_2^1(0, \ell) \mid f(0) = -\Theta(\ell)\}.$$

Since $H = c\hbar D_\Theta$, with $\Theta = -1$, by Theorem 17.4 (see (17.10) in Case (ii) with $\Theta = -1$) we have

$$\lim_{n \to \infty} |(e^{-it/(n\hbar)H}\phi, \phi)|^{2n} = \lim_{n \to \infty} |(e^{-ict/nD_{-1}}\phi, \phi)|^{2n}$$

$$= e^{-|-\phi(\ell)+\phi(0)|^2 c|t|} = e^{-c|\Delta\phi|^2|t|},$$

which proves (20.2).   □

More generally, just repeating the proof presented above for an arbitrary Θ, $|\Theta| = 1$, we have the following

Corollary 20.2. *Let* H^Φ, $\Phi \in \mathbb{R}$, *be the self-adjoint realization of the differential expression*

$$H^\Phi = ic\hbar\frac{d}{dx}$$

on

$$\mathrm{Dom}(H^\Phi) = \{f \in W_2^1((0, \ell)) \mid f(0) = e^{-i\Phi}f(\ell)\}.$$

Then,

$$\lim_{n \to \infty} |(e^{-it/(\hbar n)H^\Phi}\phi, \phi)|^{2n} = e^{-\tau_\Phi|t|}, \tag{20.4}$$

where

$$\tau_\Phi = c|\Delta_\Phi\phi|^2 = c|e^{-i\Phi}\phi(\ell) - \phi(0)|^2. \tag{20.5}$$

In particular, if

$$e^{-i\Phi}\phi(\ell) \neq \phi(0),$$

then ϕ is a resonant state under continuous monitoring of the unitary evolution

$$\phi \mapsto e^{-it/\hbar H^\Phi}\phi$$

governed by the Hamiltonian H^Φ.

Remark 20.3. Notice that the magnetic Hamiltonian H^Φ is unitarily equivalent to the operator $H + \mathfrak{e}\mathcal{A}(x)$ with

$$\Phi = \frac{\mathfrak{e}}{c\hbar} \int_0^\ell \mathcal{A}(x)dx.$$

Here $\mathfrak{e}$ is the "charge" of the ♭ぃ-particle, $\mathcal{A}(x)$ is the magnetic potential (we assume that $\mathcal{A}(x)$ is a piecewise real-valued continuous function), and $\int_0^\ell \mathcal{A}(x)dx$ is the flux of the field through the ring.

Indeed, denote by U the unitary multiplication operator

$$(U\Psi)(x) = \exp\left[i\frac{\mathfrak{e}}{c\hbar} \int_0^x \mathcal{A}(s)ds\right] \cdot \Psi(x), \quad \Psi \in L^2((0,\ell)).$$

Then

$$U^*(H + \mathfrak{e}\mathcal{A}(x))U = H^\Phi,$$

which follows from the equality

$$\left(ic\hbar\frac{d}{dx} + \mathfrak{e}\mathcal{A}(x)\right)[\mathcal{E}(x) \cdot \Psi(x)] = \mathcal{E}(x) \cdot ic\hbar\frac{d}{dx}\Psi(x),$$

where

$$\mathcal{E}(x) = \exp\left[i\frac{\mathfrak{e}}{c\hbar} \int_0^x \mathcal{A}(s)ds\right],$$

and the observation that

$$\mathrm{Dom}(H) = U(\mathrm{Dom}(H^\Phi)) = U\left(\{f \in W_2^1((0,\ell)) \mid f(0) = e^{-i\Phi}f(\ell)\}\right).$$

Theorem 20.1 and Corollary 20.2 clearly suggest that continuous observation over a quantum system should rather be treated in the framework of open quantum systems theory. Below is a suitable model for that.

Theorem 20.4. *Given* $\Phi \in [0, 2\pi)$, *in the Hilbert space* $\mathcal{H} = L^2((0,\ell)) \oplus \mathbb{C}$ *introduce the maximal dissipative operator* $\widehat{H}^\Phi$ *defined on*

$$\mathrm{Dom}(\widehat{H}^\Phi) = \left\{ \begin{pmatrix} f \\ c \end{pmatrix} \bigg| f \in W_2^1((0,\ell)), \ c = f(0) \right\}$$

as

$$\widehat{H}^\Phi \begin{pmatrix} f \\ c \end{pmatrix} = i \begin{pmatrix} \frac{d}{dx}f(x) \\ f(0) - e^{-i\Phi}f(\ell) \end{pmatrix}.$$

If $\phi \in \mathcal{H}$ *is a state such that* $\phi \in \mathrm{Dom}(\widehat{H}^\Phi)$,

$$\int_0^\ell |\phi(x)|^2 dx + |\phi(0)|^2 = 1,$$

then

$$\lim_{n \to \infty} |(e^{(it/n)\widehat{H}^{\Phi}} \phi, \phi)|^{2n} = e^{-\tau_{\Phi}|t|}, \tag{20.6}$$

where

$$\tau_{\Phi} = |e^{-i\Phi}\phi(\ell) - \phi(0)|^2. \tag{20.7}$$

Proof. We have

$$(\widehat{H}^{\Phi}f, f) = i \int_0^{\ell} f'(x)\overline{f(x)}dx + i(f(0) - e^{-i\Phi}f(\ell))\overline{f(0)}, \quad f \in \text{Dom}(\widehat{H}^{\Phi}).$$

In particular,

$$\text{Im}\left((\widehat{H}^{\Phi}f, f)\right) = \frac{1}{2}|f(\ell)|^2 - \frac{1}{2}|f(0)|^2 + |f(0)|^2 - \text{Re}\left(e^{-i\Phi}f(\ell)\overline{f(0)}\right)$$

$$= \frac{1}{2}|e^{-i\Phi}f(\ell) - f(0)|^2 \geq 0,$$

which shows that $\widehat{H}^{\Phi}$ is a dissipative operator. To complete the proof it remains to check that the lower half-plane belongs to the resolvent set of $\widehat{H}^{\Phi}$, so that $\widehat{H}^{\Phi}$ is a maximal dissipative operator, and then use the same reasoning as the one in the proof of Lemma 15.15.   □

Remark 20.5. The idea to add to the original Hilbert space $\mathcal{H}$ a one-dimensional "vacuum" subspace $\mathbb{C}$ is due to Schrader [123], also see [84], [103] and [126], where such extensions of a non-densely-defined symmetric operators found applications in modeling three-body systems with δ-like interactions that are free of the "fall to the center" phenomenon. For the general extension theory for non-densely-defined operators and its applications we also refer to [4, 61, 63, 102].

20.1. Discussion

The decay law (20.2) shows that continuous monitoring eventually triggers an exponential decay of the system. Meanwhile, the explicit expression (20.3) for the decay constant τ, $\tau = c|\Delta\psi|^2$, suggests that we are dealing with an interference alternative, which means that we cannot apply the laws of probabilities (18.1) and have to count on the composition laws of amplitudes (18.2). Indeed, a particle arriving at the junction point, the point of observation, has two options: a) either to stay on the track or b) be emitted. However, there is no way to "experimentally" confirm which option has been realized in reality. That is, we are not certain about what happened at the junction point $x = 0 = \ell$, and consequently, the description of motion becomes an interference alternative.

On the quantitative level, the reasoning presented above can be supported by the following considerations.

The incoming $\phi(\ell)$, outgoing $\phi(0)$ and emission amplitudes ϕ_{em} are to satisfy the "interference" relation

$$\phi(\ell) = \phi_{em} + \phi(0). \tag{20.8}$$

Since the quantity $c|\phi_{em}|^2 t$ asymptotically describes the probability that the emitted particle can eventually be detected during the time interval $[0, t]$,[1] the probability $P(t)$ of staying on the ring should fall off exponentially as

$$P(t) = e^{-c|\phi_{em}|^2 t} = e^{-c|\Delta\phi|^2 t}, \quad t \geq 0, \tag{20.10}$$

---

[1]This can be justified as follows. Suppose we put in an ideal detector that counts particles passing through the point x_0 in the interval $(0, \ell)$. If the initial state ϕ is a smooth function in a neighborhood of x_0, the probability $p_{x_0}(t)$ that the detector will detect a particle during the time interval t is asymptotically given by

$$p_{x_0}(t) = c|\phi(x_0)|^2 t + o(t) \quad \text{as} \quad t \to 0. \tag{20.9}$$

To justify the claim, recall that in accordance with the probabilistic interpretation of the wave-function, the probability to find the particle inside the interval $\delta \subset [0, \ell]$ is given by

$$\Pr\{\text{"particle"} \in \delta\} = \int_\delta |\phi(x)|^2 dx.$$

In particular,

$$\Pr\{\text{"particle"} \in [x_0 - \varepsilon, x_0]\} = \int_{x_0 - \varepsilon}^{x_0} |\phi(x)|^2 dx$$

$$= |\phi(x_0)|^2 \varepsilon + o(\varepsilon) \quad \text{as} \quad \varepsilon \to 0.$$

If we repeat the experiment N_∞ times, the quantity

$$\Delta N = N_\infty \int_{x_0 - \varepsilon}^{x_0} |\phi(x)|^2 dx$$

gives the (average) number of outcomes when the (quantum) particle is accommodated by the interval $[x_0 - \varepsilon, x_0]$.

One can change the point of view and assume that we are dealing with a beam of particles and that initially there were N_∞ particles in the system. Therefore, ΔN would have the meaning of the averaged number of particles in the interval $[x_0 - \varepsilon, x_0]$. For the quantum system in question wave-particle duality is exact and hence we may assume that the particles are moving to the right with speed of light c. Therefore, in time

$$t = \frac{\varepsilon}{c}$$

all the particles will leave the interval $[x_0 - ct, x_0]$ and will eventually be counted by the detector.

where

$$|\phi_{\text{em}}|^2 = |\Delta\phi|^2 = |\phi(\ell) - \phi(0)|^2, \tag{20.11}$$

which gives a heuristic justification of decay law (20.2) in Theorem 20.1.

Notice that if $\phi \in \text{Dom}(H)$, then the wave function is continuous at the junction point, that is,

$$\phi(\ell) = \phi(0). \tag{20.12}$$

Therefore, $|\phi_{\text{em}}|^2 = 0$ by (20.11) and hence there is no emission of particles. In this case, the dynamics is frozen by the continuous monitoring and we face the quantum Zeno effect.

However, in the situation in question, in view of Corollary 20.2 and Remark 20.3 one can unfreeze the evolution (stopped by continuous monitoring) by switching on the magnetic field through the ring. Indeed, since the configuration space of the system (the ring S) is not a simply connected set, the effect of the magnetic potential will be to produce the phase shift of the wave function [2] at the junction point even if the magnetic field is absent in a neighborhood of the ring S (the Aharonov-Bohm effect)

$$\phi(\ell) \mapsto \phi(\ell)e^{-i\Phi},$$

where

$$\Phi = \frac{\mathfrak{e}}{c\hbar} \int_0^\ell \mathcal{A}(x)dx.$$

Here $\mathfrak{e}$ is the "charge" of the ♭-particle and $\int_0^\ell \mathcal{A}(x)dx$ is the flux of the field through the ring.

In this case, the interference relation (20.8) should be modified as

$$e^{-i\Phi}\phi(\ell) = \phi_{\text{em}} + \phi(0).$$

Therefore, the decay properties of the state under continuous monitoring are determined by the quantity

$$|\Delta_\Phi\phi|^2 = |e^{-i\Phi}\phi(\ell) - \phi(0)|^2, \tag{20.13}$$

Finally, the probability that the detector will go off within the time interval $[0, t]$ is asymptotically given by

$$p_{x_0}(t) \sim \frac{\Delta N}{N_\infty} = \int_{x_0 - ct}^{x_0} |\phi(x)|^2 dx = c|\phi(x_0)|^2 t + o(t) \quad \text{as} \quad t \to 0,$$

which justifies the claim (20.9).

which finally leads to the decay law

$$P(t) = e^{-|\phi_{\mathrm{em}}|^2 ct} = e^{-|\Delta_\Phi \phi|^2 ct}, \quad t \geq 0,$$

for the probability $P(t)$ to detect a particle remaining on the ring.

In particular, under the assumption that $|\phi(\ell)| = |\phi(0)|$, it follows from (20.13) that the decrement $|\Delta_\Phi \phi|^2$ experiences quite typical Aharonov-Bohn oscillations which are periodic with respect to the flux of the field. That is,

$$|\Delta_\Phi \phi|^2 = |e^{-i\Phi} e^{i \arg \phi(\ell)} - e^{i \arg \phi(0)}|^2 \cdot |\phi(\ell)|^2$$

$$= 4 \sin^2 \left(\frac{\Phi - \Delta \arg \phi}{2} \right) \cdot |\phi(\ell)|^2, \tag{20.14}$$

where

$$\Delta \arg \phi = \arg \phi(\ell) - \arg \phi(0).$$

The above discussion provides a physically motivated example of a quantum system with the decay law (20.4), (20.5), see Corollary 20.2.

Chapter 21

CONTINUOUS MONITORING WITH NO INTERFERENCE

Continuous monitoring of open quantum systems leads to a completely different understanding of decay processes.

We will assume that the time evolution of the system is governed by a semi-group of contractive transformations generated by a dissipative differentiation operator such that the initial state belongs to the domain of the operator.

Theorem 21.1. *Suppose that* $\mathbb{Y} = (0, \ell)$ *is a metric graph in Case* (ii). *Given* $|\varkappa| < 1$, *in the Hilbert space* $\mathcal{H} = L^2(\mathbb{Y})$ *denote by* $\widehat{H}_\varkappa$ *the dissipative differentiation operator*

$$\widehat{H}_\varkappa = ic\hbar \frac{d}{dx}$$

on

$$\mathrm{Dom}(\widehat{H}_\varkappa) = \left\{ f \in W_2^1((0, \ell)) \,|\, f(0) = \varkappa f(\ell) \right\}. \tag{21.1}$$

Assume that $\phi \in \mathrm{Dom}(\widehat{H}_\varkappa)$ *and* $\|\phi\| = 1$.
Then

$$\lim_{n \to \infty} |(e^{it/(\hbar n)\widehat{H}_\varkappa} \phi, \phi)|^{2n} = e^{-\tau t}, \quad t \geq 0, \tag{21.2}$$

where

$$\tau = c\Delta|\phi|^2 = c(|\phi(\ell)|^2 - |\phi(0)|^2). \tag{21.3}$$

Proof. Integration by parts yields

$$\int_0^\ell \phi'(x)\overline{\phi(x)}dx = |\phi(\ell)|^2 - |\phi(0)|^2 - \int_0^\ell \phi(x)\overline{\phi'(x)}dx,$$

and therefore

$$\mathrm{Re}\int_0^\ell \phi'(x)\overline{\phi(x)}dx = \frac{\int_0^\ell \phi'(x)\overline{\phi(x)}dx + \overline{\int_0^\ell \phi'(x)\overline{\phi(x)}dx}}{2}$$

$$= \frac{|\phi(\ell)|^2 - |\phi(0)|^2}{2} \ge 0 \quad (\text{for } \phi \in \mathrm{Dom}(\widehat{H}_\varkappa)).$$

Since $\phi \in \mathrm{Dom}(\widehat{H}_\varkappa)$, as in the proof of Lemma 15.15 one obtains

$$\lim_{n\to\infty} |(e^{it/(n\hbar)\widehat{H}_\varkappa}\phi, \phi)|^{2n} = e^{-2\hbar^{-1}\mathrm{Im}(\widehat{H}_\varkappa\phi,\phi)t}, \quad t \ge 0,$$

with

$$2\hbar^{-1}\mathrm{Im}(\widehat{H}_\varkappa\phi, \phi)t = 2c\,\mathrm{Im}\left(i\int_0^\ell \phi'(x)\overline{\phi(x)}dx\right)$$

$$= c(|\phi(\ell)|^2 - |\phi(0)|^2) = \tau. \qquad \square$$

One can go back from the reduced description of the open quantum system to the full one following the extended Hilbert space approach presented below.

Consider an open quantum system prepared in the state $\phi \in L^2(\mathbb{S}) = L^2((0, \ell))$ the time evolution of which generated by the dissipative operator $\widehat{H}_\varkappa$ with the boundary condition parameter $\varkappa$, $|\varkappa| < 1$, referred to in Theorem 21.1.

Suppose that $\phi \in W_2^1((0, \ell))$ is such that the radiation condition

$$|\phi(\ell)| > |\phi(0)| \tag{21.4}$$

holds. Assume, in addition, that $\phi \in \mathrm{Dom}(\widehat{H}_\varkappa)$, that is,

$$\varkappa = \frac{\phi(0)}{\phi(\ell)}.$$

Along with the open quantum system in the state space $L^2((0, \ell))$ introduce a new quantum system in an extended Hilbert space $\mathfrak{H}$ containing $L^2((0, \ell))$ as a (proper) subspace. For the Hamiltonian $\mathbb{H}$ of the new system in the extended Hilbert space we choose a (minimal) self-adjoint dilation of the dissipative operator $\widehat{H}_\varkappa$ and the new state of the system $\hat{\phi} \in \mathfrak{H}$ is a clone of ϕ considered as an element of the extended Hilbert space $\mathfrak{H}$.

The results of continuous monitoring of the quantum evolution $\widehat{\phi} \mapsto e^{-it/\hbar \mathsf{H}} \widehat{\phi}$ generated by the self-adjoint Hamiltonian H in the Hilbert space $\mathfrak{H}$ can be described as follows.

Corollary 21.2. *Suppose that Υ is the metric graph $\Upsilon = (0, \ell)$ in Case (ii). Given $|\varkappa| < 1$, in the Hilbert space $\mathcal{H} = L^2(\Upsilon)$ denote by $\widehat{H}_\varkappa$ the dissipative differentiation operator*

$$\widehat{H}_\varkappa = ic\hbar \frac{d}{dx}$$

on

$$\mathrm{Dom}(\widehat{H}_\varkappa) = \left\{ f \in W_2^1((0, \ell)) \mid f(0) = \varkappa f(\ell) \right\}. \tag{21.5}$$

Let H be a self-adjoint dilation of the dissipative operator $\widehat{H}_\varkappa$ in an extended Hilbert $\mathfrak{H}$ space containing the original Hilbert space $\mathcal{H}$ as a (proper) subspace $\mathcal{H} = L^2((0, \ell)) \subset \mathfrak{H}$. Suppose that

$$\phi \in \mathcal{H}, \quad \|\phi\| = 1.$$

Then

$$\lim_{n \to \infty} |(e^{-it/(\hbar n)\mathsf{H}} \phi, \phi)|^{2n} = \lim_{n \to \infty} |(e^{i|t|/(\hbar n)\widehat{H}_\varkappa} \phi, \phi)|^{2n}, \quad t \in \mathbb{R},$$

provided that at least one (and therefore both) of the limits exist.

In particular, if

$$\phi \in \mathrm{Dom}(\widehat{H}_\varkappa),$$

then

$$\lim_{n \to \infty} |(e^{-it/(\hbar n)\mathsf{H}} \phi, \phi)|^{2n} = e^{-\tau t}, \quad t > 0,$$

where the decay constant τ is given by

$$\tau = c(|\phi(\ell)|^2 - |\phi(0)|^2). \tag{21.6}$$

21.1. Discussion

The decay law (21.2), (21.3) suggests that the interference effects are definitely absent. In order to get an adequate explanation for the phenomenon, instead of applying the composition law of amplitudes (20.8) one has to use the calculus of probabilities

$$|\phi(\ell)|^2 = |\phi_{\mathrm{em}}|^2 + |\phi(0)|^2. \tag{21.7}$$

Here is an argument supporting (21.7): a particle arriving at the junction point still has two options: a) either to stay on the track or b) be emitted. However, the transition from the open quantum system with state space $\mathcal{H} = L^2(\mathbb{S})$ referred to in Theorem (21.1) to the closed one in the extended Hilbert space $\mathfrak{H}$ containing $\mathcal{H}$ as a proper subspace and discussed in Corollary 21.2 assumes that an additional scattering channel $\mathfrak{H} \ominus \mathcal{H}$ is added and the emitted particles as well as the ones stayed on the track can eventually be counted. In other words, we are dealing with an exclusive alternative.

From the experimental viewpoint in this case, the arrangement of the corresponding Gedankenexperiment involves the installation of two additional detectors D_ℓ and D_0 that count the particles that pass through the point $x = \ell$ and $x = 0$. In other words, we accept the experimental condition that the emitted particles can eventually be counted (by combining the readings of the two detectors.)

Given (21.7), arguing as in Chapter 19 we arrive to the exponential decay law

$$P(t) = e^{-|\phi_{\mathrm{em}}|^2 ct} = e^{-c\Delta|\phi|^2 t}, \quad t \geq 0, \tag{21.8}$$

where $P(t)$ stands for the probability to detect the particle on the ring at the moment of time t and

$$|\phi_{\mathrm{em}}|^2 = \Delta|\phi|^2 = |\phi(\ell)|^2 - |\phi(0)|^2. \tag{21.9}$$

Below we offer an heuristic explanation of the law (21.9) based on purely classical interpretation of the nature of a גל-particle.

When we watch the beam of גל-particles by observing the readings of the two detectors, we indeed deal with an exclusive alternative. Denote by $N_\infty(t)$ the total amount of particles on the ring at the moment of time t and let N_ℓ be the amount of particles that passed through the point $x = \ell$ and arrived at the check point $x = 0$ during the time interval $[0, t]$. The arrived particles "have" the alternative: either to keep moving on the ring or to be emitted. By checking the readings of the detector D_0 we know that N_0 out of N_∞ particles stayed traveling along the ring. Next, taking into account the readings of the detector D_ℓ, we conclude that the remaining $\Delta N = N_\ell - N_0$ particles have beed emitted during the time interval $[0, t]$. (Here we implicitly assume that there is no other mechanism that causes the particles to radiate. Why this hypothesis is consistent with the way of the suggested reasoning will be explained later.)

Given the wave-function probabilistic interpretation above, it is easy to see that

$$\frac{N_\ell}{N_\infty} = c|\phi(\ell-)|^2 t + o(t) \quad \text{and} \quad \frac{N_0}{N_\infty} = c|\phi(0+)|^2 t + o(t) \quad \text{as} \quad t \to 0.$$

Therefore,

$$\frac{\Delta N}{N_\infty} = \left(|\phi(\ell)|^2 - |\phi(0)|^2\right) ct + o(t).$$

Repeating that monitoring over and over, in the limit $t \to 0$, we arrive at the differential equation

$$\frac{dN}{dt} = -c\left(|\phi(\ell)|^2 - |\phi(0)|^2\right) N,$$

$$N(0) = N_\infty,$$

that governs the counting process.

Therefore, under continuous monitoring with detectors that are going off, the total number of particles $N(t)$ as a function of time falls off exponentially as

$$N(t) = N_\infty e^{-c\Delta|\phi|^2 t}, \tag{21.10}$$

where the decrement $\Delta|\phi|^2 > 0$ is given by (21.9) (provided that $|\phi|$ has a jump at the point $x = 0$).

Notice that if the state ϕ is a continuous function on the ring, then no emission is observed and then the quantum Zeno effect takes place.

Summarizing, we arrive at the conclusion that the computation of the emission probability for the quantum system (H, ϕ) referred to in Theorem 20.1 requires the application of the composition law of amplitudes (20.8) (the interference alternative scenario). Meanwhile, the decay rate for the open quantum system $(\widehat{H}_\varkappa, \phi)$ referred to in Theorem 21.1 or for the quantum system $(\mathbb{H}, \widehat{\phi})$ in the extended Hilbert space $\mathfrak{H}$ (see Corollary 23.3) can be evaluated using the rules of the calculus of probabilities (21.7) (the exclusive alternative scenario).

Chapter 22

THE SELF-ADJOINT DILATION

The self-adjoint dilation $\mathbb{H}$ of the dissipative operator $\widehat{H}_{\varkappa}$ referred to in Corollary 21.2 can be described explicitly as the differentiation operator on the metric graph $\mathbb{Y}$ with appropriate boundary conditions. However, the geometry of the metric graph that determines the configuration space of the quantum system depends on whether or not the parameter $\varkappa$ in the boundary condition (21.1) vanishes.

If $\varkappa \neq 0$, as it follows from Theorem 5.7, the extended Hilbert space can be chosen to coincide with $\mathfrak{H} = L^2(\mathbb{Y}) = L^2(\overline{\mathbb{Y}})$ where $\mathbb{Y}$ is the metric graph

$$\mathbb{Y} = (-\infty, 0) \sqcup (0, \ell) \sqcup (0, \infty) \quad \text{in Case } (iii)$$

and $\overline{\mathbb{Y}}$ denotes its one-cycle completion, the configuration space of the extended quantum system, while the self-adjoint dilation $\mathbb{H}$, the Hamiltonian, coincides with the differentiation operator on the graph

$$\mathbb{H} = ic\hbar \frac{d}{dx} \tag{22.1}$$

defined on the domain of functions satisfying the boundary conditions

$$\begin{pmatrix} f_\infty(0+) \\ f_\ell(0) \end{pmatrix} = \begin{pmatrix} |\varkappa| & -\sqrt{1 - |\varkappa|^2}\,\frac{\varkappa}{|\varkappa|} \\ \sqrt{1 - |\varkappa|^2} & \varkappa \end{pmatrix} \begin{pmatrix} f_\infty(0-) \\ f_\ell(\ell) \end{pmatrix}. \tag{22.2}$$

In the exceptional case $\varkappa = 0$, the boundary condition (21.1) that determines the dissipative operator $\widehat{H}_0$ is local. Therefore, it is convenient to assume that the configuration space for the corresponding open quantum system (if $\varkappa = 0$) is the finite interval $(0, \ell)$ rather than a ring. Consequently, in this case the extended Hilbert space may be chosen as $\mathfrak{H} = L^2(\mathbb{Y})$

where $\mathbb{Y}$ is the metric graph

$$\mathbb{Y} = (-\infty, 0) \sqcup (0, \ell) \sqcup (\ell, \infty) \quad \text{in Case } (i).$$

The self-adjoint dilation (the Hamiltonian) $\mathbb{H}$ of the dissipative operator $\widehat{H}_0$ can be chosen to be the differentiation operator on the graph

$$\mathbb{H} = ic\hbar\frac{d}{dx} \tag{22.3}$$

with the self-adjoint boundary condition

$$f(\ell+) = \Theta f(\ell-), \quad |\Theta| = 1, \tag{22.4}$$

with an arbitrary choice of the unimodular extension parameter Θ.

Notice that in the limit $\varkappa \to 0$ along the ray $\varkappa = -|\varkappa|\Theta$ the boundary conditions (22.2) split as

$$f_\infty(0+) = \Theta f_\ell(\ell)$$

and

$$f_\infty(0-) = f_\ell(0). \tag{22.5}$$

In view of (22.5), the one-cycle graph $\overline{\mathbb{Y}}$ "unwinds" to a straight line

$$\mathbb{Y} = (-\infty, 0) \sqcup (0, \ell) \sqcup (\ell, \infty)$$

which can naturally be identified with the real axis. One can show that the corresponding self-adjoint dilations of the dissipative operators $\widehat{H}_{-|\varkappa|\Theta}$ approach (in the strong resolvent sense) the operator (22.3) with the boundary condition (22.4).

Notice that emission amplitude ϕ_{em} from (21.9) can be evaluated by directly solving the Schrödinger equation

$$i\hbar\frac{\partial}{\partial t}\Psi = \mathbb{H}\Psi \tag{22.6}$$

on the one-cycle graph $\overline{\mathbb{Y}}$.

Indeed, representing the initial state $\widehat{\phi}$ in the extended Hilbert space $\mathfrak{H} = L^2(\overline{\mathbb{Y}})$ as the two-component vector function

$$\widehat{\phi} = \begin{pmatrix} \widehat{\phi}_\infty \\ \widehat{\phi}_\ell \end{pmatrix} = \begin{pmatrix} 0 \\ \phi \end{pmatrix}, \quad \widehat{\phi} \in L^2(\overline{\mathbb{Y}}),$$

denoted by $\Psi(t,x)$, $x \in \overline{Y}$ the solution of the Schrödinger equation (22.6) with the initial condition

$$\Psi|_{t=0} = \widehat{\phi} = \begin{pmatrix} 0 \\ \phi \end{pmatrix}.$$

We claim that emission amplitude ϕ_{em} can be evaluated as the limiting value of the first component $\Psi_\infty(t, 0+)$ of the solution Ψ as $t \to 0$.

Indeed, since

$$\lim_{t \downarrow 0} \lim_{\varepsilon \downarrow 0} \Psi(t, \varepsilon) = \begin{pmatrix} f_\infty(0+) \\ f_\ell(0) \end{pmatrix} = \begin{pmatrix} |\varkappa| & -\sqrt{1 - |\varkappa|^2}\frac{\varkappa}{|\varkappa|} \\ \sqrt{1 - |\varkappa|^2} & \varkappa \end{pmatrix} \begin{pmatrix} 0 \\ \phi(\ell) \end{pmatrix},$$

we have

$$\phi_{\mathrm{em}} = \Psi_\infty(0+, 0+) = -\sqrt{1 - |\varkappa|^2}\frac{\varkappa}{|\varkappa|}\phi(\ell).$$

Therefore, the probability density of the event of emission of a particle is given by

$$|\phi_{\mathrm{em}}|^2 = (1 - |\varkappa|^2)|\phi(\ell)|^2 = |\phi(\ell)|^2 - |\phi(0)|^2 = \Delta|\phi|^2,$$

which agrees with (21.2), (21.3) (cf. Remark 17.7). Here we have used that the boundary condition

$$\phi(0) = \varkappa\phi(\ell)$$

holds, that is,

$$\phi \in \mathrm{Dom}(\widehat{H}_\varkappa).$$

Notice, that in the exceptional case, that is, if the initial state is such that $\phi(0) = 0$, the dissipative operator $\widehat{H}_0$ associated with the corresponding open quantum system has the domain

$$\mathrm{Dom}(\widehat{H}_0) = \{f \in W_2^1((0, \ell)) \mid f(0) = 0\}.$$

In this case, the differentiation operator $\mathbb{H}$ on the whole real axis,

$$\mathbb{H} = ic\hbar\frac{d}{dx} \quad \text{on } \mathrm{Dom}(\mathbb{H}) = W_2^1((-\infty, \infty)),$$

dilates $\widehat{H}_0$. Therefore, the configuration space of the extended quantum system is just the real axis $Y = \mathbb{R}$, not the one-cycle graph $\overline{Y}$. The corresponding solution $\Psi(x, t)$ of the Schrödinger equation (22.6) with the initial data

$$\Psi|_{t=0} = \phi$$

(here we do not distinguish the function ϕ on $[0, \ell]$ and its extension by zero on the whole real axis $\mathbb{R}$) is a one-component function given by

$$\Psi(x, t) = \phi(x - ct).$$

In this case the emission amplitude ϕ_{em} can be evaluated as

$$\phi_{\text{em}} = \Phi(x - ct)|_{x=\ell,\, t=0} = \phi(\ell),$$

so that again

$$|\phi_{\text{em}}|^2 = |\phi(\ell)|^2 = |\phi(\ell)|^2 - |\phi(0)|^2 = \Delta|\phi|^2.$$

It is also worth mentioning that the decrement $\Delta|\phi|^2$ given by (21.9) and referred to in Theorem 21.1 is gauge invariant while $|\Delta\phi|^2$ defined in (20.11) is not.

Indeed, if

$$(Vf)(x) = e^{i\lambda(x)} f(x), \quad x \in (0, \ell), \tag{22.7}$$

is a (unitary) gauge transformation, where $\lambda(x)$ is a differentiable function on $[0, \ell]$, then

$$\Delta|V\phi|^2 = \Delta|\phi|^2$$

and

$$|\Delta V\phi|^2 = |e^{i\Phi}\phi(\ell) - \phi(0)|^2 \neq |\Delta\phi|^2 \quad \text{(in general)}.$$

Here,

$$\Phi = \lambda(\ell) - \lambda(0) = \int_0^\ell \frac{d}{dx}\lambda(x)dx.$$

is a shift of the relative phase of the wave function.

As we have already pointed out, if the wave function is continuous at the junction point, that is

$$\phi(0) = \phi(\ell),$$

then

$$\Delta|\phi|^2 = |\Delta\phi|^2 = 0.$$

In this case the quantum Zeno effect takes place regardless of whether both of the detectors D_0 and D_ℓ go off or only one of them does. Moreover, if V is a gauge transformation, we also have that

$$\Delta |V\phi|^2 = \Delta |\phi|^2 = 0$$

that shows that the Zeno effect is stable with respect to the gauge transformations when both of the detectors D_0 and D_ℓ go off. The situation is quite different in the experiment when only one of the detectors goes off. A tiny variation of the phase of the wave functions can easily transfer the system from the quantum Zeno mode to the exponential decay regime with the decay rate given by the "magnetic" decrement

$$|\Delta_\Phi \phi|^2 = |e^{-i\Phi}\phi(\ell) - \phi(0)|^2.$$

We remark that

$$(|\phi(\ell)| - |\phi(0)|)^2 \le |\Delta_\Phi \phi|^2 \le (|\phi(\ell)| + |\phi(0)|)^2.$$

Moreover, by changing the "flux" Φ, the upper and as well as the lower bound for the magnetic decrement can easily be attained. In particular,

$$0 \le |\Delta_\Phi \phi|^2 \le 4|\phi(0)|^2,$$

whenever the continuity condition (20.12) at the junction point holds.

Chapter 23

GENERAL OPEN QUANTUM SYSTEMS ON A RING

Notice that while considering open quantum systems $(\widehat{H}_\varkappa, \phi)$ on a ring referred to in Theorem 21.1, we assumed

$$\phi \in \text{Dom}(\widehat{H}_\varkappa).$$

This requirement can be relaxed and we arrive at the following more general result.

Theorem 23.1. *Suppose that* $\phi \in W_2^1((0,\ell))$ *is a state,* $\|\phi\| = 1$. *Given* $|\varkappa| \le 1$, *in the Hilbert space* $\mathcal{H} = L^2(\mathbb{Y})$, $\mathbb{Y} = (0,\ell)$, *denote by*

$$\widehat{H}_\varkappa = i c \hbar \frac{d}{dx}$$

the differentiation operator with the boundary condition

$$f(0) = \varkappa f(\ell),$$

that is,

$$\text{Dom}(\widehat{H}_\varkappa) = \left\{ f \in W_2^1((0,\ell)) \mid f(0) = \varkappa f(\ell) \right\}.$$

Then

$$\lim_{n \to \infty} |(e^{it/n \widehat{H}_\varkappa} \phi, \phi)|^{2n} = e^{-\tau t}, \quad t > 0, \tag{23.1}$$

where

$$\tau = |\phi(0) - \varkappa \phi(\ell)|^2 + (1 - |\varkappa|^2)|\phi(\ell)|^2. \tag{23.2}$$

Proof. Without loss we will assume that we work in the system of units where $c = 1$ and $\hbar = 1$.

It is sufficient to prove the asymtotic representation

$$\text{Re}(e^{it\widehat{H}_\varkappa}\phi, \phi) = 1 - \frac{1}{2}\tau t + o(t) \quad \text{as} \quad t \downarrow 0. \tag{23.3}$$

Denote by $W(t) = e^{it\widehat{H}_\varkappa}$ the contractive semi-group generated by the operator $\widehat{H}_\varkappa$. Notice that

$$W(t + \ell) = \varkappa W(t), \quad t \geq 0,$$

with

$$(W(t)\phi(x) = \begin{cases} \varkappa\phi(\ell + x - t), & 0 < x < t \\ \phi(x - t), & x < t < \ell \end{cases}.$$

Assume that $\phi \in W_2^1((0, \ell))$ is the (absolutely) continuous representative of the element ϕ (denoted by the same symbol). We have

$$(W(t)\phi, \phi) = \int_0^t \varkappa\phi(\ell + x - t)\overline{\phi(x)}dx + \int_t^\ell \phi(x - t)\overline{\phi(x)}dx = I + J,$$

where

$$I = \varkappa \int_0^t \phi(\ell + x - t)\overline{\phi(x)}dx$$

and

$$J = \int_t^\ell \phi(x - t)\overline{\phi(x)}dx.$$

Since ϕ is a continuous function, we get

$$I = t\varkappa\phi(\ell)\overline{\phi(0)} + o(t) \quad \text{as} \quad t \downarrow 0.$$

Moreover, the membership $\phi \in W_2^1((0, \ell))$ ensures the representation

$$\phi(x - t) = \phi(x) - t\phi'(x) + \eta_t(x), \quad \text{for a.e.} \quad x \in [\ell - t, \ell]$$

where ϕ' denotes the generalized derivative of ϕ and

$$\|\eta_t\|_{L^2((\ell-t,\ell))} = o(t) \quad \text{as} \quad t \downarrow 0.$$

Therefore,

$$
\begin{aligned}
J &= \int_t^\ell \phi(x)\overline{\phi(x)}dx - t\int_t^\ell \phi(x)\overline{\phi'(x)}dx + o(t)\\
&= \int_0^\ell \phi(x)\overline{\phi(x)}dx - \int_0^t \phi(x)\overline{\phi(x)}dx - t\int_0^\ell \phi(x)\overline{\phi'(x)}dx + o(t)\\
&= \int_0^\ell \phi(x)\overline{\phi(x)}dx - t|\phi(0)|^2 - t\int_0^\ell \phi(x)\overline{\phi'(x)}dx + o(t)\\
&= 1 - t\left[|\phi(0)|^2 + \int_0^\ell \phi(x)\overline{\phi'(x)}dx\right] + o(t) \quad \text{as} \quad t\downarrow 0.
\end{aligned}
$$

Here we have used that ϕ is a state, that is,

$$
\|\phi\|^2 = \int_0^\ell \phi(x)\overline{\phi(x)}dx = 1.
$$

Combining the asymptotic representations for I and J we get

$$
I + J = 1 - t\left[|\phi(0)|^2 + \int_0^\ell \phi(x)\overline{\phi'(x)}dx - \varkappa\phi(\ell)\overline{\phi(0)}\right] + o(t) \quad \text{as} \quad t\downarrow 0.
$$

Since

$$
\mathrm{Re}\left(\int_0^\ell \phi(x)\overline{\phi(x)}dx\right) = \frac{|\phi(\ell)|^2 - |\phi(0)|^2}{2},
$$

we have

$$
\begin{aligned}
\mathrm{Re}(e^{it\widehat{H}_\varkappa}\phi, \phi) = \mathrm{Re}(W(t)\phi, \phi) &= \mathrm{Re}(I + J)\\
&= 1 - \frac{1}{2}\tau t + o(t) \quad \text{as} \quad t\downarrow 0.
\end{aligned}
$$

Here

$$
\begin{aligned}
\tau &= |\phi(0)|^2 + |\phi(\ell)|^2 - 2\mathrm{Re}\left(\varkappa\phi(\ell)\overline{\phi(0)}\right)\\
&= |\phi(0) - \varkappa\phi(\ell)|^2 + (1 - |\varkappa|^2)|\phi(\ell)|^2,
\end{aligned}
$$

which proves (23.3). $\square$

Remark 23.2. Notice that Theorems 20.1 and 21.1 (with $c = \hbar = 1$) are particular cases of the obtained result. Indeed, if $|\varkappa| = 1$ and therefore the operator $\widehat{H}_{\varkappa}$ is self-adjoint, then the expression for the decay rate τ in (23.2) simplifies to

$$\tau = |\phi(0) - \varkappa\phi(\ell)|^2 = |\Delta_{\varkappa}\phi|^2$$

(cf. Theorem 20.1 ($\varkappa = 1$)).

On the other hand, if

$$\phi \in \mathrm{Dom}(\widehat{H}_{\varkappa}),$$

we have that $\phi(0) = \varkappa\phi(\ell)$ and therefore the general expression (23.2) for the decay rate reduces to

$$\tau = (1 - |\varkappa|^2)|\phi(\ell)|^2 = |\phi(\ell)|^2 - |\phi(0)|^2 = \Delta|\phi|^2 \qquad (23.4)$$

(cf. Theorem 21.1).

Applying Corollary 21.2 and taking into account that the self-adjoint dilation of the dissipative operator $\widehat{H}_{\varkappa}$ is now explicitly available, see Chapter 21, in view of Theorem 23.1, we arrive at the following two results. It is convenient to threat the case $\varkappa \neq 0$ and the exceptional case $\varkappa = 0$ separately.

Theorem 23.3. ($\varkappa \neq 0$) *Suppose that* $\mathbb{Y} = (-\infty, 0) \sqcup (0, \ell) \sqcup (0, \infty)$ *is a metric graph in Case (iii). Given* $0 < |\varkappa| < 1$*, in the Hilbert space* $\mathcal{H} = L^2(\mathbb{Y})$ *denote by* $\mathbb{H}$ *the differentiation operator*

$$\mathbb{H} = ic\hbar\frac{d}{dx} \qquad (23.5)$$

defined on the domain of functions satisfying the boundary conditions

$$\begin{pmatrix} f_\infty(0+) \\ f_\ell(0) \end{pmatrix} = \begin{pmatrix} |\varkappa| & -\sqrt{1 - |\varkappa|^2}\frac{\varkappa}{|\varkappa|} \\ \sqrt{1 - |\varkappa|^2} & \varkappa \end{pmatrix} \begin{pmatrix} f_\infty(0-) \\ f_\ell(\ell) \end{pmatrix}.$$

Suppose that $\widehat{\phi} = \begin{pmatrix} 0 \\ \phi \end{pmatrix} \in \mathfrak{H}$ *is a state* ($\|\phi\| = 1$) *such that*

$$\phi \in W_2^1((0, \ell)) \subset L^2(\mathbb{Y}). \qquad (23.6)$$

Then

$$\lim_{n \to \infty} |(e^{-it/(\hbar n)\mathbb{H}}\widehat{\phi}, \widehat{\phi})|^{2n} = e^{-\tau|t|}, \quad t \in \mathbb{R}, \qquad (23.7)$$

where

$$\tau = c|\phi(0) - \varkappa\phi(\ell)|^2 + c(1 - |\varkappa|^2)|\phi(\ell)|^2. \qquad (23.8)$$

Theorem 23.4. ($\varkappa = 0$) *Suppose that* $\Upsilon = (-\infty, 0) \sqcup (0, \ell) \sqcup (\ell, \infty)$ *is a metric graph in Case* (*i*). *In the Hilbert space* $\mathcal{H} = L^2(\Upsilon)$ *denote by* $\mathbb{H}$ *the differentiation operator*

$$\mathbb{H} = i c \hbar \frac{d}{dx} \tag{23.9}$$

defined on $W_2^1(\Upsilon)$.
Suppose that $\phi \in W_2^1((0, \ell)) \subset L^2(\Upsilon)$ *and* $\|\phi\| = 1$.
Then

$$\lim_{n \to \infty} |(e^{-it/(\hbar n)\mathbb{H}} \phi, \phi)|^{2n} = e^{-\tau|t|}, \quad t \in \mathbb{R}, \tag{23.10}$$

where

$$\tau = c(|\phi(0)|^2 + |\phi(\ell)|^2). \tag{23.11}$$

Below we provide an independent (illustrative) proof of Theorem 23.4 based on a direct application of the Gnedenko-Kolmorogov limit theorem.

Proof. Let $\widehat{\phi}$ denote the Fourier transform of the state ϕ,

$$\widehat{\phi}(\lambda) = \frac{1}{\sqrt{2\pi}} \int_{-\infty}^{\infty} e^{-i\lambda x} \phi(x) dx = \frac{1}{\sqrt{2\pi}} \int_{0}^{\ell} e^{-i\lambda x} \phi(x) dx.$$

Under the hypotheses of $\phi \in W_2^1((0, \ell))$, we integrate by parts and obtain the asymptotic representation

$$\widehat{\phi}(\lambda) = \frac{i}{\sqrt{2\pi}} \frac{\phi(\ell) e^{-i\ell\lambda} - \phi(0)}{\lambda} + o\left(\frac{1}{|\lambda|}\right) \quad \text{as} \quad |\lambda| \to \infty.$$

We have

$$\int_{\lambda}^{\infty} |\widehat{\phi}(s)|^2 ds = \frac{1}{2\pi} \int_{\lambda}^{\infty} \frac{|\phi(\ell)|^2 + |\phi(0)|^2}{s^2} ds$$

$$- 2\mathrm{Re}\left[\phi(\ell)\overline{\phi(0)} \int_{\lambda}^{\infty} e^{-i\ell s} \frac{ds}{s^2}\right] + o\left(\frac{1}{|\lambda|}\right)$$

$$= \frac{|\phi(\ell)|^2 + |\phi(0)|^2}{2\pi} \frac{1}{\lambda} + o\left(\frac{1}{|\lambda|}\right) \quad \text{as} \quad \lambda \to \infty.$$

In a completely similar way one shows that

$$\int_{-\infty}^{\lambda} |\widehat{\phi}(s)|^2 ds = \frac{|\phi(\ell)|^2 + |\phi(0)|^2}{2\pi} \frac{1}{|\lambda|} + o\left(\frac{1}{|\lambda|}\right) \quad \text{as} \quad \lambda \to -\infty.$$

Therefore, the distribution function $F(x)$ of the (absolutely continuous) probability measure $|\widehat{\phi}(x)|^2 dx$ satisfies the hypotheses (H.4) and (H.5) of

Theorem H.1 in Appendix H with $\alpha = 1$, $h(x) = 1$ and

$$c_1 = c_2 = \frac{|\phi(\ell)|^2 + |\phi(0)|^2}{2\pi} \cdot \frac{\pi}{2} = \frac{|\phi(\ell)|^2 + |\phi(0)|^2}{4}.$$

Therefore, the law $F(x)$ belongs to the normal domain of attraction of the symmetric 1-stable law the characteristic function of which is given by

$$e^{-\frac{1}{2}|(|\phi(\ell)|^2 + |\phi(0)|^2)||t|}.$$

In particular,

$$\lim_{n \to \infty} |\Phi(t/n)|^{2n} = e^{-|(|\phi(\ell)|^2 + |\phi(0)|^2)||t|},$$

where

$$\Phi(t) = \int_{\mathbb{R}} e^{itx} dF(x)$$

is the characteristic function of the probability distribution $|\widehat{\phi}(x)|^2 dx$. Since

$$(e^{-it/\hbar H}\phi, \phi) = \Phi(-ct),$$

we conclude that

$$\lim_{n \to \infty} |(e^{-it/(\hbar n)H}\phi, \phi)|^{2n} = \lim_{n \to \infty} |\Phi(-ct/n)|^{2n} = e^{-\tau|t|},$$

where τ is given by (23.11). □

The main idea of the proof (with appropriate minimal adjustments) can be used to obtain the following more general result.

Suppose that the state ϕ is a piecewise continuous function with discontinuity points $a_1 < a_2 < \ldots, a_N$ such that

$$\phi \in W_2^1((-\infty, a_1)) \oplus W_2^1((a_1, a_2)) \oplus \cdots \oplus W_2^1((a_{N-1}, a_N)) \oplus W_2^1((a_N, \infty)).$$

Then

$$\lim_{n \to \infty} |(e^{-it/(\hbar n)H}\phi, \phi)|^{2n} = e^{-\tau|t|},$$

where

$$\tau = c \sum_{k=1}^{N} |\Delta\phi(a_k)|^2$$

and

$$\Delta\phi(a_k) = \phi(a_k + 0) - \phi(a_k - 0), \quad k = 1, 2, \ldots, N,$$

is the jump of the piecewise continuous representative of the state ϕ at the point a_k.

23.1. Discussion

The decay law (23.1) shows that the decay rate (23.2) splits into two terms,

$$\tau = \tau_{\text{excl}} + \tau_{\text{inter}},$$

where

$$\tau_{\text{excl}} = c(1 - |\varkappa|^2)|\phi(\ell)|^2 \tag{23.12}$$

and

$$\tau_{\text{inter}} = c|\phi(0) - \varkappa\phi(\ell)|^2. \tag{23.13}$$

Recall that the configuration space of the open quantum system referred to in Theorem 23.1 is the ring $\mathcal{S}$ obtained from the interval $[0, \ell]$ by identifying its end-points. As the result of gluing the ends of the interval, a "point defect" occurs which can be perceived as a kind of membrane which can be characterized by a quantum gate coefficient $\varkappa$, the amplitude that the particle goes through the membrane. Arguing as a pure probabilist, one can conclude that the particle penetrates through the membrane with probability $|\varkappa\phi(\ell)|^2$ and therefore with probability proportional to $(1 - |\varkappa|^2)|\phi(\ell)|^2$ it should be emitted. In other words, the law of probabilities is applicable

$$|\phi(\ell)|^2 = |\phi_{\text{em}}^{\text{excl}}|^2 + |\varkappa|^2|\phi(\ell)|^2. \tag{23.14}$$

However, a secondary emission mechanism is available: the amplitude $\varkappa\phi(\ell)$ that the particle can be found to the right from the membrane interferes with the amplitude $\phi(0)$ to stay on the ring $\mathcal{S}$. So composition law for amplitudes

$$\varkappa\phi(\ell) = \phi_{\text{em}}^{\text{inter}} + \phi(0) \tag{23.15}$$

should be take into account.

It remains to recall that the emission amplitudes and the corresponding decay constants are related as

$$\tau_{\text{excl}} = c|\phi_{\text{em}}^{\text{excl}}|^2 \tag{23.16}$$

and

$$\tau_{\text{inter}} = c|\phi_{\text{em}}^{\text{inter}}|^2 \tag{23.17}$$

and then (23.12) and (23.13) follow from (23.14), (23.16) and (23.15), (23.17), respectively.

Summarizing, we arrive at the following descriptive understanding of the decay processes under continuous monitoring. A בּ-particle moves along the ring from 0 to ℓ as a particle, hits the membrane and with some probability is emitted following the classical scheme (21.9) discussed in Chapter 20. After the collision with the membrane, the particle transforms into a wave, interferes with itself and experiences the secondary emission following the radiation szenario described in Chapter 19.

The suggested interpretation allows one to enhance the status of informal reasoning that led to the decay law (21.10): set

$$\varkappa = \frac{\phi(0)}{\phi(\ell)}$$

and observe that

$$\tau_{\text{inter}} = |\phi(0) - \varkappa\phi(\ell)|^2 = 0$$

whenever the initial state ϕ belongs to the domain of the dissipative operator $\widehat{H}_\varkappa$. This is what was implicitly assumed in the derivation of the law (21.10). Having these remarks in mind, one can consider that derivation being an informal retelling of the rigorous time-dependent proof of Theorem 23.1 in the case where the initial state ϕ belongs to the domain of $\widehat{H}_\varkappa$.

23.2. Random Phase Method

The traditional way of deriving the laws of probabilities (18.1) from the law of amplitudes (18.2), (18.3) is based on the hypothesis that "the performance of the corresponding experiment will necessarily alter the phase" [44] of the wave function by an unknown amount which eventually, after averaging, yields the law of probabilities (18.1).

We suggest the following mnemonic rule for heuristic derivation of the decay law when dealing with the open system referred to in Theorem 23.1.

Start with the Kirchhoff rule for the amplitudes (see (20.8))

$$\phi_{\text{em}} = \phi(\ell) - \phi(0)$$

and rewrite the rule in the following equivalent form

$$\phi_{\text{em}} + \varkappa\phi(\ell) = \phi(\ell) + (\varkappa\phi(\ell) - \phi(0)).$$

The performance of the corresponding experiment assumes counting incoming and emitted particles by observing readings of the corresponding detectors.

Rewrite as

$$\Theta_{em}\phi_{em} + \varkappa\phi(\ell) = \Theta_\ell\phi(\ell) + (\varkappa\phi(\ell) - \phi(0)),$$

where Θ_{em} and Θ_ℓ are unimodular independent random variables with zero mean. After the corresponding averaging we get

$$\mathsf{E}|\Theta_{em}\phi_{em} + \varkappa\phi(\ell)|^2 = \mathsf{E}|\Theta_\ell\phi(\ell) + (\varkappa\phi(\ell) - \phi(0))|^2,$$

where E denotes the corresponding mathematical expectation. Since Θ_{em} and Θ_ℓ are independent with zero mean, we arrive at the law

$$|\phi_{em}|^2 + |\varkappa|^2\phi(\ell)|^2 = |\phi(\ell)|^2 + |\varkappa\phi(\ell) - \phi(0)|^2.$$

Hence

$$|\phi_{em}|^2 = (1 - |\varkappa|^2)|\phi(\ell)|^2 + |\varkappa\phi(\ell) - \phi(0)|^2,$$

which coincides with the decay rate τ given by (23.2).

Chapter 24

OPERATOR COUPLING
LIMIT THEOREMS

We say that a dissipative operator $\widehat{A}$ belongs to class $\mathfrak{D}(\mathcal{H})$ if $\widehat{A}$ is a quasi-selfadjoint extension of a symmetric operators $\dot{A}$ with deficiency indices $(1,1)$ (see Appendix G). Recall that in this case the symmetric operator $\dot{A}$ can be recovered from $\widehat{A}$ as

$$\dot{A} = \widehat{A}\big|_{\mathrm{Dom}(\widehat{A}) \cap \mathrm{Dom}((\widehat{A}))^*}. \tag{24.1}$$

Definition 24.1. We will say that two dissipative operators $\widehat{A}_1 \in \mathfrak{D}(\mathcal{H}_1)$ and $\widehat{A}_2 \in \mathfrak{D}(\mathcal{H}_2)$ coincide in distribution (are equally distributed), in writing

$$\widehat{A}_1 \overset{\mathrm{d}}{=} \widehat{A}_2$$

if there are appropriate self-adjoint reference operators $A_{1,2}$ such that the characteristic functions of the triples $(\dot{A}_1, \widehat{A}_1, A_1)$ and $(\dot{A}_2, \widehat{A}_2, A_2)$ coincide. That is,

$$S_{(\dot{A}_1, \widehat{A}_1, A_1)}(z) = S_{(\dot{A}_2, \widehat{A}_2, A_2)}(z), \quad z \in \mathbb{C}_+.$$

Here the symmetric operators $\dot{A}_{1,2}$ ($\dot{A}_{1,2} \subset A_{1,2}$) are defined by (24.1).

Recall that an operator $\widehat{A}$ in the Hilbert space $\mathcal{H}$ is completely non-self-adjoint if $\mathcal{H}$ cannot be represented in the form of an orthogonal sum of two subspaces $\mathcal{H}_1$ and $\mathcal{H}_0 \neq 0$ with the following properties, see, e.g., [15]:

1) $\mathcal{H}_1$ and $\mathcal{H}_0$ are invariant relative to $\widehat{A}$;
2) $\widehat{A}$ induces in $\mathcal{H}_0$ a self-adjoint operator.

It is worth mentioning that if completely non-selfadjoint operators $\widehat{A}_1 \in$ $\mathfrak{D}(\mathcal{H}_1)$ and $\widehat{A}_2 \in \mathfrak{D}(\mathcal{H}_2)$ are equally distributed, then they are unitarily equivalent. Indeed, in this case the corresponding symmetric operators $\dot{A}_{1,2}$ are prime and then the claim follows from the Uniqueness Theorem C.1 in Appendix C.

Definition 24.2. We say that a sequence of dissipative operators $\{\widehat{A}_n\}_{n=1}^{\infty}$, $\widehat{A}_n \in \mathfrak{D}(\mathcal{H}_n)$ converges in distribution to a dissipative operator $\widehat{A} \in \mathfrak{D}(\mathcal{H})$, in writing,

$$\lim_{n\to\infty} \widehat{A} \stackrel{\mathrm{d}}{=} \widehat{A},$$

if one can find a sequence of reference self-adjoint operators $A_n = A_n^*$ in $\mathcal{H}_n$ and $A = A^*$ in $\mathcal{H}$ such that the corresponding characteristic functions converge pointwise. That is,

$$\lim_{n\to\infty} S_{(\dot{A}_n, \widehat{A}_n, A_n)}(z) = S_{(\dot{A}, \widehat{A}, A)}(z), \quad z \in \mathbb{C}_+.$$

As long as convergence in distribution for a sequence of operators is introduced, the next natural question is to understand the behavior of the n-fold coupling of an operator with itself as $n \to \infty$.

The following limit theorem sheds some light on the typical behavior of n-fold couplings of operators from the class $\mathfrak{D}(\mathcal{H})$.

Theorem 24.3. *Suppose that $\widehat{A}$ is a maximal dissipative operator from the class $\mathfrak{D}(\mathcal{H})$. Assume that the characteristic function $S(z) = S_{(\dot{A}, \widehat{A}, A)}(z)$ associated with the triple $(\dot{A}, \widehat{A}, A)$ for some (and therefore with all) reference self-adjoint extension A admits an analytic continuation to the lower half-plane in a neighborhood of a point $\mu \in \mathbb{R}$.*

(i) *If $|S(\mu)| = 1$, then*

$$\lim_{n\to\infty} n \underbrace{(\widehat{A} - \mu I) \uplus (\widehat{A} - \mu I) \uplus \cdots \uplus (\widehat{A} - \mu I)}_{n \text{ times}} \stackrel{\mathrm{d}}{=} \widehat{D}_{II}(\ell),$$

where $\widehat{D}_{II}(\ell)$ is the dissipative differentiation operator on the finite interval $[0, \ell]$ with

$$\mathrm{Dom}(\widehat{D}_{II}(\ell)) = \{f \in W_2^1((0, \ell)) \,|\, f(0) = 0\}$$

and

$$\ell = \frac{1}{i} \frac{d}{d\lambda} \log S_{(\widehat{A}, A)}(\mu) > 0. \tag{24.2}$$

(ii) *If $|S(\mu + i0)| < 1$, then*

$$\lim_{n\to\infty} n \underbrace{(\widehat{A} - \mu I) \uplus (\widehat{A} - \mu I) \uplus \cdots \uplus (\widehat{A} - \mu I)}_{n \; times} \stackrel{d}{=} \widehat{D}_I(0), \qquad (24.3)$$

where $\widehat{D}_I(0)$ is the dissipative differentiation operator on the real axis on

$$\mathrm{Dom}(\widehat{D}_I(0)) = \{f \in W_2^1((-\infty,0)) \oplus W_2^1((0,\infty)) \,|\, f(0+) = 0\}.$$

Proof. (i) Introduce the n-fold coupling

$$\widehat{B}_n = \biguplus_{k=1}^{n} n(\widehat{A} - \mu I).$$

Combining the invariance principle (see Theorem F.1 in Appendix F) and the Multiplication Theorem G.5 in Appendix G, one ensures the existence of the corresponding reference operators B_n and unimodular factors Θ_n such that

$$S_{(\dot{B}_n, \widehat{B}_n, B_n)}(z) = \Theta_n \left(S_{(\dot{A}, \widehat{A}, A)}\left(\frac{z}{n} + \mu\right)\right)^n$$

$$= \Theta_n \left(S_{(\dot{A},\widehat{A},A)}(\mu) + i\frac{d}{d\lambda}S_{(\dot{A},\widehat{A},A)}(\mu)\frac{z}{n} + o(n^{-1})\right)^n$$

$$\text{as} \quad n \to \infty.$$

By choosing a possibly different sequence of reference operators B'_n one obtains that

$$S_{(\dot{B}_n,\widehat{B}_n,B'_n)}(z) = \left(1 + i\ell\frac{z}{n} + o(n^{-1})\right)^n \quad \text{as} \quad n \to \infty,$$

where ℓ is given by (24.2).

Therefore,

$$\lim_{n\to\infty} S_{(\dot{B}_n,\widehat{B}_n,B'_n)}(z) = e^{i\ell z}, \quad z \in \mathbb{C}_+.$$

To complete the proof it remains to notice that

$$S_{(\dot{D}_{II},\widehat{D}_{II}(\ell),D_{II})}(z) = e^{i\ell z}, \quad z \in \mathbb{C}_+,$$

where $(\dot{D}_{II}, \widehat{D}_{II}(\ell), D_{II})$ is the triple $(\dot{D}, \widehat{D}(\ell), D)$ referred to in Lemma 9.1 in Case (ii).

(ii) From Theorem F.1 in Appendix F it follows that there exists a sequence of unimodular factors $|\Theta_n| = 1$ such that

$$S_{(n(\dot{A}-\mu I),(n(\widehat{A}-\mu I),n(A-\mu I))}(z) = \Theta_n S_{(\dot{A},\widehat{A},A)}\left(\frac{z}{n} + \mu\right).$$

Since

$$\lim_{n\to\infty} S_{(\dot{A},\widehat{A},A)}\left(\frac{z}{n} + \mu\right) = S(\mu + i0), \quad z \in \mathbb{C}_+,$$

one can always choose reference operators A'_n such that

$$\lim_{n\to\infty} S_{((n(\dot{A}-\mu I),n(\widehat{A}-\mu I),n(A'-\mu I))}(z) = |S(\mu + i0)| = k. \tag{24.4}$$

Since $k < 1$, we have that

$$\lim_{n\to\infty} \left(S_{(n(\dot{A}-\mu I),(n(\widehat{A}-\mu I),n(A-\mu I))}(z)\right)^n = 0 \quad \text{for all } z \in \mathbb{C}_+. \tag{24.5}$$

However, by Lemma 9.1, we have that

$$S_{(\dot{D}_I(0),\widehat{D}_I(0),D_I(0))}(z) = 0, \quad z \in \mathbb{C}_+,$$

and hence (24.5) means that the sequence of operators

$$\widehat{B}_n = \underbrace{(\widehat{A} - \mu I) \uplus (\widehat{A} - \mu I) \uplus \cdots \uplus (\widehat{A} - \mu I)}_{n \text{ times}}$$

converges to $\widehat{D}_I(0)$ in distribution. □

Remark 24.4. A closer look at the proof shows that (ii) is a simple consequence of the limit relation (combine (24.4) and Lemma 9.1)

$$\lim_{n\to\infty} (n(\widehat{A} - \mu)I) \overset{\mathrm{d}}{=} \widehat{D}_I(k), \tag{24.6}$$

where $k = |S(\mu + i0)|$ and $\widehat{D}_I(k)$ is the dissipative differentiation operator on the real axis on

$$\mathrm{Dom}(\widehat{D}_I(0)) = \{f \in W_2^1((-\infty,0)) \oplus W_2^1((0,\infty)) \mid f(0+) = kf(0-)\}.$$

Remark 24.5. Borrowing the terminology from probability theory, we will say that an operator $\widehat{A}$ has a (strictly) stable distribution if for arbitrary constants c_1 and c_2 there exists a constant c such that

$$c_1 \widehat{A}_1 \uplus c_2 \widehat{A}_2 \overset{\mathrm{d}}{=} c\widehat{A},$$

whenever

$$\widehat{A}_1 \overset{\mathrm{d}}{=} \widehat{A}_2 \overset{\mathrm{d}}{=} \widehat{A}.$$

In particular, the operator $\widehat{D}_{II}(\ell)$ has a stable distribution since

$$c_1 \widehat{D}_{II}(\ell) \uplus c_2 \widehat{D}_{II}(\ell) \overset{\mathrm{d}}{=} c\widehat{D}_{II}(\ell) \tag{24.7}$$

with

$$\frac{1}{c} = \frac{1}{c_1} + \frac{1}{c_2} \tag{24.8}$$

The stability laws (24.7) and (24.8) in particular imply

$$n \cdot \underbrace{\widehat{D}_{II}(\ell) \uplus \widehat{D}_{II}(\ell) \uplus \cdots \uplus \widehat{D}_{II}(\ell)}_{n \text{ times}} \overset{\mathrm{d}}{=} \widehat{D}_{II}(\ell) \tag{24.9}$$

and

$$\frac{1}{n} \cdot \underbrace{V_\ell \uplus V_\ell \uplus \cdots \uplus V_\ell}_{n \text{ times}} \overset{\mathrm{d}}{=} V_\ell, \tag{24.10}$$

which is in a good agreement with limit Theorems 24.3 and Theorem 24.6 (see below).

Although the distribution laws for the operators $\widehat{D}_I(k)$, $0 < k < 1$, and $\widehat{D}_{III}(k, \ell)$ are not stable, nevertheless corresponding laws are infinitely divisible in the sense that

$$n \cdot \underbrace{\widehat{D}_I(k^{1/n}) \uplus \widehat{D}_I(k^{1/n}) \uplus \cdots \uplus \widehat{D}_I(k^{1/n})}_{n \text{ times}} \overset{\mathrm{d}}{=} \widehat{D}_I(k),$$

where $\widehat{D}_I(k)$ is the dissipative differentiation operator on the real axis on

$$\mathrm{Dom}(\widehat{D}_I(k)) = \{f \in W_2^1((-\infty, 0)) \oplus W_2^1((0, \infty)) \mid f(0+) = kf(0-)\},$$

and

$$n \cdot \underbrace{\widehat{D}_{III}\left(k^{1/n}, n^{-1}\ell\right) \uplus \cdots \uplus \widehat{D}_{III}\left(k^{1/n}, n^{-1}\ell\right)}_{n \text{ times}} \overset{\mathrm{d}}{=} \widehat{D}_{III}(k, \ell),$$

with $\widehat{D}_{III}(k, \ell)$ the dissipative differentiation operator on the metric graph

$$\mathbb{Y} = (-\infty, 0] \sqcup [0, \infty) \sqcup [0, \ell]$$

on

$$\text{Dom}(\widehat{D}_{III}(k, \ell)) = \left\{ f_\infty \oplus f_\ell \in W_2^1(\mathbb{Y}) \mid \begin{cases} f_\infty(0+) & = k f_\infty(0-) \\ f_\ell(0+) & = \sqrt{1 - k^2} f_\infty(0-) \end{cases} \right\}.$$

It is also worth mentioning that by Lemma 9.1 and the Multiplication Theorem G.5 in Appendix G, the basic differentiation operators $\widehat{D}_I(k)$, $\widehat{D}_{II}(\ell)$ and $\widehat{D}_{III}(k, \ell)$ have the "addition" laws with respect to the operator coupling

$$\widehat{D}_I(k) \uplus \widehat{D}_I(k') \overset{\text{d}}{=} \widehat{D}_I(kk'), \tag{24.11}$$

while

$$\widehat{D}_{II}(\ell) \uplus \widehat{D}_{II}(\ell') \overset{\text{d}}{=} \widehat{D}_{II}(\ell + \ell'). \tag{24.12}$$

One also gets the following spectral synthesis rule

$$\widehat{D}_{III}(k, \ell) \uplus \widehat{D}_{III}(k', \ell') \overset{\text{d}}{=} \widehat{D}_{III}(kk', \ell + \ell'). \tag{24.13}$$

Notice that in the first two cases one can even state that the equalities in distribution imply the unitary equivalence of the corresponding operators.

The Limit Theorem 24.3 (i) has its natural counterpart for the class $\mathfrak{D}_0(\mathcal{H})$ (see Appendix G for the definition of the class).

Theorem 24.6. *Suppose that $\widehat{A} \in \mathfrak{D}_0(\mathcal{H})$ is a maximal bounded dissipative operator. Then*

$$\lim_{n \to \infty} \frac{1}{n} \underbrace{\widehat{A} \uplus \widehat{A} \uplus \cdots \uplus \widehat{A}}_{n \text{ times}} \overset{d}{=} V_\ell, \tag{24.14}$$

where V_ℓ is the Volterra operator in $L^2((0, \ell))$

$$(V_\ell f)(x) = i \int_0^x f(t)dt, \quad f \in L^2((0, \ell)),$$

and

$$\ell = 2\text{tr}\left(\text{Im}(\widehat{A})\right).$$

Proof. Since

$$\left(\frac{1}{2i}(V_\ell - V_\ell^*)f\right)(x) = \frac{1}{2}\int_0^\ell f(t)dt \cdot \chi_{[0,\ell]}(x), \quad f \in L^2((0,\ell)),$$

where

$$\chi_{[0,\ell]}(x) = 1 \quad \text{on} \quad [0,\ell],$$

the imaginary part of V_ℓ is one-dimensional and therefore $V_\ell \in \mathfrak{D}_0(L^2((0,\ell)))$.

Next, we evaluate the characteristic function of the (bounded) operator V_ℓ (see (A.10) in Appendix A). We have

$$S_{V_\ell}(z) = 1 + i(V_\ell^* - zI)^{-1}\chi_{[0,\ell]}, \chi_{[0,\ell]}).$$

Since

$$((V_\ell^* - zI)^{-1}\chi_{[0,\ell]})(x) = \frac{(-1)}{z}e^{\frac{i}{z}(x-\ell)}, \quad x \in [0,\ell],$$

we finally get that

$$S_{V_\ell}(z) = 1 - \frac{1}{iz}\int_0^\ell e^{\frac{i}{z}(x-\ell)}dx = \exp\left(-i\frac{\ell}{z}\right), \quad z \in \mathbb{C}_+.$$

Next, since $\widehat{A} \in \mathfrak{D}_0(\mathcal{H})$, we have the representation

$$\widehat{A} = A + i(\cdot g)g$$

for some $g \in \mathcal{H}$, $g \neq 0$. Therefore, the characteristic function $S_{\widehat{A}}$ has the following asymptotics

$$S_{\widehat{A}}(z) = 1 + 2i((\widehat{A})^* - zI)^{-1}g, g) = 1 - 2i\frac{\text{tr}(\text{Im}(\widehat{A}))}{z} + o\left(\frac{1}{z}\right) \quad \text{as} \quad z \to \infty.$$

Set

$$B_n = \frac{\widehat{A} \uplus \widehat{A} \uplus \cdots \uplus \widehat{A}}{n}.$$

Using the invariance principle, Theorem F.1 in Appendix F, and the Multiplication Theorem G.2 in Appendix G, one concludes that

$$\lim_{n \to \infty} S_{B_n}(z) = \lim_{n \to \infty} S_{\widehat{A}}(nz) = \lim_{n \to \infty}\left(1 - 2i\frac{\text{tr}(\text{Im}(\widehat{A}))}{nz} + o\left(\frac{1}{n}\right)\right)^n$$

$$= \exp\left(-\frac{i\ell}{z}\right), \quad z \in \mathbb{C}_+.$$

Therefore, $\lim_{n \to \infty} S_{B_n}(z)$ coincides with the characteristic function of the Volterra operator V_ℓ which completes the proof. □

Remark 24.7. Notice that the addition laws with respect to operator coupling (24.11)–(24.13) can be extended by the following rule

$$V_\ell \uplus V_{\ell'} \overset{d}{=} V_{\ell+\ell'},$$

which can be deduced from (24.12) applying the invariance principe combined with the observation that

$$V_\ell = -D_{II}(\ell)^{-1}.$$

In particular, the operator V_ℓ has a stable distribution:

$$(c_1' V_\ell) \uplus (c_2' V_\ell) \overset{d}{=} c' V_\ell,$$

with

$$c' = c_1' + c_2'.$$

We conclude this chapter by the discussion of the limit distribution *universality* of the real part of n-fold couplings (as $n \to \infty$) of bounded dissipative operators from the class $\mathfrak{D}_0(\mathcal{H})$.

Corollary 24.8. *Suppose that $\widehat{A} \in \mathfrak{D}_0(\mathcal{H})$ is a maximal bounded dissipative operator and $\ell = 2\mathrm{tr}\left(\mathrm{Im}(\widehat{A})\right)$. Let*

$$B_n = \frac{1}{n} \underbrace{\widehat{A} \uplus \widehat{A} \uplus \cdots \uplus \widehat{A}}_{n \ times}$$

denote the "averaged" n-fold coupling of the operator $\widehat{A}$ with itself.

Let $M_n(z)$ be the Weyl-Titschmarsh function of the self-adjoint operator $\mathrm{Re}(\widehat{B}_n)$,

$$M_n(z) = \frac{1}{\mathrm{tr}(\mathrm{Im}(\widehat{B}_n))} \mathrm{tr}\left((\mathrm{Re}(\widehat{B}_n) - zI)^{-1} \mathrm{Im}(\widehat{B}_n)\right).$$

Suppose that $\mu_n(d\lambda)$ is the probability measure from the representation theorem

$$M_n(z) = \int_{\mathbb{R}} \frac{d\mu_n(\lambda)}{\lambda - z}, \quad z \in \mathbb{C}_+.$$

Then the sequence of the measure $\{\mu_n(d\lambda)\}_{n=1}^{\infty}$ converges weakly to the pure point probability measure $\mu(d\lambda)$ given by

$$\mu(d\lambda) = \frac{4}{\pi^2} \sum_{k \in \mathbb{Z}} \frac{1}{(2k+1)^2} \delta_{z_k}(d\lambda), \qquad (24.15)$$

with the "Dirac masses" $\delta_{z_k}(d\lambda)$ at the points

$$z_k = \frac{1}{\ell} \frac{2}{(k+\frac{1}{2})\pi}, \quad k \in \mathbb{Z}.$$

That is,

$$\lim_{n \to \infty} \mu_n((-\infty, \lambda)) = \mu((-\infty, \lambda))$$

at every point of continuity of the function $F(\lambda) = \mu((-\infty, \lambda))$.

Proof. From Lemma A.4 in Appendix A it follows that

$$M_n(z) = \frac{1}{it_n} \frac{S_{\widehat{B}_n}(z) - 1}{S_{\widehat{B}_n}(z) + 1},$$

where $S_{\widehat{B}_n}(z)$ is the characteristic function of $\widehat{B}_n$ and

$$t_n = \operatorname{tr}(\operatorname{Im}(\widehat{B}_n)) = \operatorname{tr}(\operatorname{Im}(\widehat{A})) = \frac{\ell}{2}.$$

Applying Theorem 24.6 we get that

$$\lim_{n \to \infty} S_{\widehat{B}_n}(z) = \exp\left(-\frac{i\ell}{z}\right)$$

and therefore

$$\lim_{n \to \infty} M_n(z) = \frac{2}{i\ell} \frac{\exp\left(-\frac{i\ell}{z}\right) - 1}{\exp\left(-\frac{i\ell}{z}\right) + 1} = -\frac{2}{\ell} \tan\frac{\ell}{2}z, \quad z \in \mathbb{C}_+.$$

Since

$$\tan(z) = -2 \sum_{k=0}^{\infty} \left(\frac{z}{z^2 - (k+\frac{1}{2})^2\pi^2}\right)$$

and hence

$$-\tan\frac{1}{z} = 2 \sum_{k=0}^{\infty} \left(\frac{z}{\left(1 - \left(k+\frac{1}{2}\right)\pi z\right)\left(1 + \left(k+\frac{1}{2}\right)\pi z\right)}\right),$$

it is easy to see that

$$-\frac{2}{\ell} \tan\frac{\ell}{2}z = \int_{\mathbb{R}} \frac{d\mu(\lambda)}{\lambda - z}, \quad z \in \mathbb{C}_+,$$

with $\mu(d\lambda)$ given by (24.15). Since μ_n and μ are probability measures and

$$\lim_{n \to \infty} M_n(z) = \lim_{n \to \infty} \int_{\mathbb{R}} \frac{d\mu_n(\lambda)}{\lambda - z} = \int_{\mathbb{R}} \frac{d\mu(\lambda)}{\lambda - z}, \quad z \in \mathbb{C}_+, \tag{24.16}$$

the pointwise convergence of the Stieltjes transforms (24.16) ensures the weak convergence of μ_n to μ by an analog of the Lévy continuity theorem [40]. $\qquad\square$

Appendix A

THE CHARACTERISTIC FUNCTION FOR RANK-ONE PERTURBATIONS

In this chapter we introduce a characteristic function of a maximal dissipative operator $\widehat{A}$ in the case where the domains of the operator and its adjoint coincide [4, 15, 68, 77, 81, 86, 100, 128], that is,

$$\mathrm{Dom}(\widehat{A}) = \mathrm{Dom}((\widehat{A})^*). \tag{A.1}$$

For instance, if the operator $\widehat{A}$ is bounded, condition (A.1) holds automatically.

We will treat the simplest case where $\widehat{A}$ has a rank-one imaginary part of the form

$$\mathrm{Im}(\widehat{A}) = \frac{1}{2i}(\widehat{A} - (\widehat{A})^*) = tP,$$

where $t > 0$ and P is a rank-one orthogonal projection,

$$P = (\cdot, g)g, \quad \|g\| = 1.$$

Denote by A the real part of $\widehat{A}$,

$$A = \mathrm{Re}(\widehat{A}) = \frac{1}{2}(\widehat{A} + (\widehat{A})^*), \quad \mathrm{Dom}(A) = \mathrm{Dom}(\widehat{A}) = \mathrm{Dom}((\widehat{A})^*),$$

so that

$$\widehat{A} = A + itP.$$

The resolvent of $\widehat{A}$ is described in the following lemma.

Lemma A.1 (cf. [124]). *Let $\widehat{A}$ be a maximal dissipative operator with a rank-one imaginary part,*

$$\widehat{A} = A + itP,$$

$A = A^*$, $P = (\cdot, g)g$, $\|g\| = 1$, and $t > 0$.
  Denote by

$$M(z) = ((A - zI)^{-1}g, g), \quad z \in \rho(A),$$

the M-function associated with the real part $\mathrm{Re}(\widehat{A})$ *of the operator* $\widehat{A}$ *and the unit vector g.*
  Then the following resolvent formula

$$(\widehat{A} - zI)^{-1} = (A - zI)^{-1} - p(z)(A - zI)^{-1}P(A - zI)^{-1},$$

$$z \in \rho(\widehat{A}) \cap \rho(A), \tag{A.2}$$

holds, where

$$p(z) = \frac{1}{M(z) + \frac{1}{it}}.$$

In particular,

$$((\widehat{A} - zI)^{-1}g, g) = \frac{M(z)}{1 + itM(z)}. \tag{A.3}$$

Moreover,

$$\mathrm{spec}(\widehat{A}) \subset \{z : 0 \leq \mathrm{Im}(z) \leq t\}. \tag{A.4}$$

Proof. To prove the resolvent formula (A.2), one observes that

$$(\widehat{A} - zI)^{-1} = (A - zI)^{-1} - it(\widehat{A} - zI)^{-1}P(A - zI)^{-1}, \tag{A.5}$$

and hence

$$(\widehat{A} - zI)^{-1}g = (A - zI)^{-1}g - it((A - zI)^{-1}g, g)(\widehat{A} - zI)^{-1}g,$$

which yields the representation

$$(\widehat{A} - zI)^{-1}g = (1 + itM(z))^{-1}(A - zI)^{-1}g.$$

Substituting this equality back to (A.5), one obtains

$$(\widehat{A} - zI)^{-1} = (A - zI)^{-1} - \frac{it}{1 + itM(z)}(A - zI)^{-1}P(A - zI)^{-1},$$

which proves (A.2).

Now, it is easy to see that the non-real spectrum of $\widehat{A}$ coincides with those z that satisfy the equation

$$\frac{1}{it} + M(z) = 0, \quad \text{Im}(z) \neq 0.$$

To complete the proof, we use the inequality

$$|M(z)| \leq \frac{1}{\text{Im}(z)}, \quad z \in \mathbb{C}_+.$$

Therefore, $\{z : \text{Im}(z) > t\} \subset \rho(\widehat{A})$, which proves (A.4). $\qquad\square$

Theorem A.2. *Let $\widehat{A}$ be a maximal dissipative operator with a rank-one imaginary part*

$$\widehat{A} = A + itP,$$

$A = A^*$, $P = (\cdot, g)g$, $\|g\| = 1$, *and* $t > 0$. *Then $\widehat{A}$ is completely non-self-adjoint if and only if the element g is generating for the self-adjoint operator* $A = \text{Re}(\widehat{A})$.

Proof. Introduce the subspace,

$$\mathcal{H}_g = \text{span}_{\delta \in \mathcal{B}(\mathbb{R})}\{E_A(\delta)g\}, \quad \text{with } \mathcal{B}(\mathbb{R}) \text{ the Borel } \sigma\text{-algebra.}$$

Only If Part. Suppose that $\widehat{A}$ is completely non-self-adjoint. Assume that g is not a generating element for the self-adjoint operator $A = \text{Re}(\widehat{A})$. Then the orthogonal complement $\mathcal{H}_g$ reduces the real part A and since the element g is orthogonal to $\mathcal{H}_g^\perp$ of $\mathcal{H}$, one obtains that

$$\widehat{A}h = Ah \quad \text{for} \quad h \in \mathcal{H}_g^\perp.$$

Since $\mathcal{H}_g$ reduces A, the subspace $\mathcal{H}_g^\perp$ reduces $\widehat{A}$ as well. Furthermore, the part of $\widehat{A}$ on $\mathcal{H}_g^\perp$ is a self-adjoint operator. Therefore, $\widehat{A}$ is not completely non-self-adjoint. We get a contradiction.

If Part. Suppose that g is a generating element for the self-adjoint operator $A = \text{Re}(\widehat{A})$. Assume that $\widehat{A}$ is not completely non-self-adjoint and let $\mathcal{H}_0$ be its reducing subspace such that the part of $\widehat{A}$ in this subspace is a self-adjoint operator.

Given $0 \neq h \in \mathcal{H}_0$, using the resolvent formula (A.2), one obtains that

$$((\widehat{A} - iyI)^{-1}h, h) = ((A - ityI)^{-1}h, h)$$
$$- \frac{((A - ityI)^{-1}h, g)((-ityI)^{-1}g, h)}{\frac{1}{it} + M(iy)},$$
$$|y| > t, \quad y \in \mathbb{R}.$$

Therefore,

$$f_1(y) = f_2(y) - \frac{f_3(y)}{f_4(y)}, \quad |y| > t, \tag{A.6}$$

where the functions f_k, $k = 1, 2, 3, 4$ are given by

$$f_1(y) = ((\widehat{A} - iyI)^{-1}h, h),$$
$$f_2(y) = ((A - ityI)^{-1}h, h),$$
$$f_3(y) = ((A - ityI)^{-1}h, g)((A - ityI)^{-1}g, h),$$
$$f_4(y) = \frac{1}{it} + M(iy).$$

One observes that

$$f_k(y) = \overline{f_k(-y)}, \quad y \in \mathbb{R}, \quad |y| > t, \quad k = 1, 2, 3. \tag{A.7}$$

However,

$$f_4(y) \neq \overline{f_4(-y)} \tag{A.8}$$

for

$$\overline{f_4(-y)} = \overline{\frac{1}{it} + M(-iy)} = \frac{(-1)}{it} + M(iy) \neq \frac{1}{it} + M(iy) = f_4(y).$$

Inequality (A.8) together with (A.7) is inconsistent with (A.6), provided that $f_3(y) \neq 0$. Finally, the fact that the function $f_3(y)$ is not identically zero easily follows from the assumption that the element g is generating for A.

The obtained contradiction shows that there is no reducing subspace such that the part of $\widehat{A}$ in this subspace is self-adjoint. Therefore, $\widehat{A}$ is completely non-self-adjoint. $\qquad\square$

Theorem A.3 (cf. [14]). *Assume that $\widehat{A}$ is a maximal dissipative operator with a rank-one imaginary part, $\widehat{A} = A + itP$, $A = A^*$, $P = (\cdot, g)g$, $\|g\| = 1$, and $t > 0$. Assume, in addition, that $\widehat{A}$ is completely non-self-adjoint.*

Let $\mu(d\lambda)$ be the probability measure from the representation

$$((A - zI)^{-1}g, g) = \int_{\mathbb{R}} \frac{d\mu(\lambda)}{\lambda - z}, \quad z \in \mathbb{C}_+.$$

Then the operator $\widehat{A}$ is unitarily equivalent to the operator $\widehat{B}$ of the form

$$(\widehat{B}f)(\lambda) = \lambda f(\lambda) + it(f, \mathbb{1})\mathbb{1}(\lambda),$$

$$\text{Dom}(\widehat{B}) = \left\{ f \in L^2(\mathbb{R}; d\mu) \,\middle|\, \int_{\mathbb{R}} \lambda^2 |f(\lambda)|^2 d\mu(\lambda) < \infty \right\},$$

in the Hilbert space $L^2(\mathbb{R}; d\mu)$, where

$$\mathbb{1}(\lambda) = 1 \quad \text{for } \mu\text{-a.e. } \lambda \in \mathbb{R}. \tag{A.9}$$

Proof. By hypothesis $\widehat{A}$ is completely non-self-adjoint and therefore, by Theorem A.2, the real part A has simple spectrum and the vector g is generating for A.

It follows that the spectral measures

$$\nu(d\lambda) = (E_A(d\lambda)g, g)$$

and

$$\mu(d\lambda) = (E_{\mathcal{B}}(d\lambda)\mathbb{1}, \mathbb{1}),$$

where $\mathcal{B}$ is the self-adjoint operator of multiplication by independent variable in $L^2(\mathbb{R}; d\mu)$ and $\mathbb{1}(\lambda)$ is given by (A.9), coincide. Since $\mathcal{B}$ has simple spectrum and $\mathbb{1}(\lambda)$ is a generating vector for the self-adjoint operator $\mathcal{B}$, the Spectral Theorem for self-adjoint operators yields the existence of a unitary operator $\mathcal{U} : \mathcal{H} \to L^2(\mathbb{R}; d\mu)$ such that

$$\mathcal{U}A\mathcal{U}^{-1} = \mathcal{B} \quad \text{and} \quad \mathcal{U}g = \mathbb{1}.$$

Hence

$$\mathcal{U}\widehat{A}\mathcal{U}^{-1} = \widehat{B},$$

which completes the proof. $\qquad\square$

Given a non-self-adjoint dissipative operator $\widehat{A}$ with a rank-one imaginary part, $\widehat{A} = A + itP$, $A = A^*$, $P = (\cdot, g)g$, $\|g\| = 1$, and $t > 0$, denote by $S(z)$ the characteristic function of operator $\widehat{A}$ following [77]

$$S(z) = 1 + 2it \left(((\widehat{A})^* - zI)^{-1}g, g \right), \quad z \in \mathbb{C}_+. \tag{A.10}$$

Lemma A.4. *Let $\widehat{A}$ be a maximal dissipative operator with a rank-one imaginary part, $\widehat{A} = A + itP$, $A = A^*$, $P = (\cdot, g)g$, $\|g\| = 1$, and $t > 0$. Then the characteristic function $S(z)$ of $\widehat{A}$ admits the representation*

$$S(z) = \frac{1 + itM(z)}{1 - itM(z)}, \quad z \in \mathbb{C}_+,$$

where $M(z)$ is the M-function of $\mathrm{Re}(\widehat{A}) = A$ given by

$$M(z) = ((A - zI)^{-1}g, g), \quad z \in \rho(A).$$

Proof. Introduce the function

$$M_t(z) = ((\widehat{A} - zI)^{-1}g, g) = ((A + itP - zI)^{-1}g, g), \quad z \in \begin{cases} \mathbb{C}_-, & \text{if } t > 0 \\ \mathbb{C}_+, & \text{if } t < 0 \end{cases}.$$

From the resolvent formula (A.2) one gets that

$$M_t(z) = \frac{M(z)}{1 + itM(z)}, \quad z \in \begin{cases} \mathbb{C}_-, & \text{if } t > 0 \\ \mathbb{C}_+, & \text{if } t < 0 \end{cases},$$

and hence

$$S(z) = 1 + 2itM_{-t}(z) = \frac{1 + itM(z)}{1 - itM(z)}, \quad z \in \mathbb{C}_+,$$

completing the proof. $\qquad\square$

Theorem A.5 ([77]). *The characteristic function of a completely non-self-adjoint operator with a rank-one imaginary part uniquely determines the operator up to unitary equivalence.*

Proof. Suppose that $\widehat{A}$ is a completely non-self-adjoint operator with a rank-one imaginary part so that

$$\widehat{A} = A + it(\cdot, g)g$$

for some $\|g\| = 1$, $A = A^*$ and $t > 0$.

In view of Theorem A.3, it suffices to show that the characteristic function uniquely determines the parameter t and the (probability) measure

μ from the representation for the M-function (in the case in question, $\mu(\mathbb{R}) = \|g\|^2 = 1$)

$$M(z) = ((A - zI)^{-1}g, g) = \int_{\mathbb{R}} \frac{d\mu(\lambda)}{\lambda - z}, \quad z \in \mathbb{C}_+. \tag{A.11}$$

Indeed, by Lemma A.4,

$$S(z) = \frac{1 + itM(z)}{1 - itM(z)} = \frac{1 - itz^{-1} + o(z^{-1})}{1 + itz^{-1} + o(z^{-1})} = 1 - \frac{2it}{z} + o(z^{-1}), \quad z \to \infty,$$

and hence

$$\frac{i}{2} \lim_{z \to \infty} z(S(z) - 1) = t,$$

which uniquely determines the perturbation parameter t. Since

$$M(z) = \frac{1}{it} \cdot \frac{S(z) - 1}{S(z) + 1},$$

the knowledge of the characteristic function $S(z)$ also uniquely determines the probability measure $\mu(d\lambda)$ in (A.11) by the Stieltjes inversion formula. $\qquad\square$

Appendix B

PRIME SYMMETRIC OPERATORS

Recall that a densely defined linear operator $\dot{A}$ in a Hilbert space $\mathcal{H}$ is called symmetric if

$$(\dot{A}x, y) = (x, \dot{A}y) \quad \text{for all} \quad x, y \in \text{Dom}(\dot{A}).$$

Definition B.1. A symmetric operator $\dot{A}$ is called a prime operator if there does not exist a (non-trivial) subspace invariant under $\dot{A}$ such that the restriction of $\dot{A}$ on this subspace is self-adjoint.

If a symmetric operator is not prime, it is useful to separate its self-adjoint part from its prime part.

Theorem B.2. *Let $\dot{A}$ be a closed symmetric operator with equal deficiency indices in a Hilbert space $\mathcal{H}$. Then the Hilbert space splits into the orthogonal sum of two subspaces*

$$\mathcal{H} = \mathcal{K} \oplus \mathcal{L}, \tag{B.1}$$

where

$$\mathcal{K} = \overline{\text{span}_{\text{Im}(z) \neq 0} \text{Ker}((\dot{A})^* - zI)}$$

and

$$\mathcal{L} = \bigcap_{\text{Im}(z) \neq 0} \text{Ran}(\dot{A} - zI).$$

Both of the subspaces $\mathcal{K}$ and $\mathcal{L}$ reduce the symmetric operator $\dot{A}$. Moreover, the part $\dot{A}|_{\mathcal{L}}$ of $\dot{A}$ in $\mathcal{L}$ is a self-adjoint operator and the part $\dot{A}|_{\mathcal{K}}$ of $\dot{A}$ in $\mathcal{K}$ is a prime symmetric operator.

Proof. Since by hypothesis the operator $\dot{A}$ is a closed symmetric operator, $\text{Ran}(\dot{A} - zI)$, $\text{Im}(z) \neq 0$, is a closed subspace and hence $\mathcal{L}$ being the intersection of closed subspaces is a closed subspace itself.

Assume that $h \in \mathcal{L}$ and hence

$$h \in \text{Ran}(\dot{A} - zI) \quad \text{for all } z \in \mathbb{C} \setminus \mathbb{R}. \tag{B.2}$$

Since

$$\mathcal{H} = \text{Ker}((\dot{A})^* - \bar{z}I) \oplus \text{Ran}(\dot{A} - zI), \quad z \in \mathbb{C} \setminus \mathbb{R}, \tag{B.3}$$

from (B.2) and (B.3) one concludes that h is orthogonal to the subspace $\text{Ker}((\dot{A})^* - zI)$ for any $z \in \mathbb{C} \setminus \mathbb{R}$. Hence h is orthogonal to $\mathcal{K}$, which means that

$$\mathcal{L} \subset \mathcal{K}^{\perp}. \tag{B.4}$$

Now, assume that an element h is orthogonal to $\mathcal{K}$.

Since the linear set $\mathcal{D} = \text{span}_{z \in \mathbb{C} \setminus \mathbb{R}} \text{Ker}((\dot{A})^* - \bar{z}I)$ is a dense subset in $\mathcal{K}$, the element h is orthogonal to $\text{Ker}((\dot{A})^* - \bar{z}I)$ for any $z \in \mathbb{C} \setminus \mathbb{R}$. Therefore, by (B.3),

$$h \in \text{Ran}(\dot{A} - zI) \quad \text{for all } z \in \mathbb{C} \setminus \mathbb{R}$$

and hence $h \in \mathcal{L}$ which means that

$$\mathcal{K}^{\perp} \subset \mathcal{L}. \tag{B.5}$$

Combining (B.4) and (B.5) completes the proof of (B.1).

To prove the remaining assertion of the theorem, we show first that the subspace $\mathcal{L}$ is $\dot{A}$-invariant.

Indeed, assume that $f \in \mathcal{L} \cap \text{Dom}(\dot{A})$ and hence

$$f \in \text{Ran}(\dot{A} - zI) \quad \text{for all } z \in \mathbb{C} \setminus \mathbb{R}. \tag{B.6}$$

Then from (B.6) follows that

$$\dot{A}f = (\dot{A} - zI)f + zf \in \text{Ran}(\dot{A} - zI) \quad \text{for all } z \in \mathbb{C} \setminus \mathbb{R},$$

and hence $\dot{A}f \in \mathcal{L}$ by the definition of the space $\mathcal{L}$.

Next, we will show that

$$(\dot{A} - zI)(\mathcal{L} \cap \text{Dom}(\dot{A})) = \mathcal{L} \quad \text{for any } z \in \mathbb{C} \setminus \mathbb{R}. \tag{B.7}$$

To see this, take an $f \in \mathcal{L}$. Then $f \in \operatorname{Ran}(\dot{A} - zI)$ for any $z \in \mathbb{C} \setminus \mathbb{R}$, and therefore, for any $z \in \mathbb{C} \setminus \mathbb{R}$ there exists an element $g_z \in \operatorname{Dom}(\dot{A})$ such that

$$f = (\dot{A} - zI)g_z.$$

We claim that

$$g_z \perp \operatorname{Ker}((\dot{A})^* - zI) \quad \text{for all } z \in \mathbb{C} \setminus \mathbb{R}.$$

Indeed, let A be a(ny) self-adjoint extension of $\dot{A}$ (recall that $\dot{A}$ has equal deficiency indices and therefore $\dot{A}$ admits self-adjoint extensions). Then

$$f = (\dot{A} - zI)g_z = (A - zI)g_z$$

and hence

$$g_z = (A - zI)^{-1}f, \quad z \in \mathbb{C} \setminus \mathbb{R}. \tag{B.8}$$

Fix a z with $\operatorname{Im}(z) \neq 0$. For any element f_ζ such that

$$f_\zeta \in \operatorname{Ker}((\dot{A})^* - \zeta I), \quad \zeta \in \mathbb{C} \setminus \mathbb{R},$$

with $\zeta \neq \bar{z}$, one gets that

$$(g_z, f_\zeta) = ((A - zI)^{-1}f, f_\zeta) = (f, (A - \bar{z}I)^{-1}f_\zeta)$$
$$= \frac{1}{z - \bar{\zeta}}\left((f, (A - \zeta I)(A - \bar{z}I)^{-1}f_\zeta) - (f, f_\zeta)\right). \tag{B.9}$$

By assumption $f \in \mathcal{L}$ and hence $f \perp \mathcal{K}$ by the first part of the proof. Since

$$(A - \zeta I)(A - \bar{z}I)^{-1}f_\zeta \in \operatorname{Ker}((\dot{A})^* - \bar{z}I) \subset \mathcal{K},$$
$$f_\zeta \in \operatorname{Ker}((\dot{A})^* - \zeta I) \subset \mathcal{K},$$

and $f \perp \mathcal{K}$, from (B.9) follows that $(g_z, f_\zeta) = 0$, i.e.,

$$g_z \perp \operatorname{Ker}((\dot{A})^* - \zeta I), \quad \zeta \neq \bar{z}, \quad \operatorname{Im}(\zeta) \neq 0. \tag{B.10}$$

It remains to show that

$$g_z \perp \operatorname{Ker}((\dot{A})^* - \bar{z}I). \tag{B.11}$$

Indeed, for any $f_{\bar{z}} \in \operatorname{Ker}((\dot{A})^* - \bar{z}I)$ we have that

$$(A - \bar{z}I)(A - \zeta I)^{-1}f_{\bar{z}} \in \operatorname{Ker}((\dot{A})^* - \zeta I).$$

Therefore, by (B.10),

$$(g_z, (A - \bar{z}I)(A - \zeta I)^{-1} f_{\bar{z}}) = 0, \quad \zeta \neq \bar{z}, \quad \mathrm{Im}(\zeta) \neq 0.$$

Hence,

$$(g_z, f_{\bar{z}}) = \lim_{\zeta \to \bar{z}} (g_z, (A - \bar{z}I)(A - \zeta I)^{-1} f_{\bar{z}}) = 0, \quad \text{for all } f_{\bar{z}} \in \mathrm{Ker}((\dot{A})^* - \bar{z}I),$$

which proves (B.11)

From (B.1) follows that $g_z \in \mathcal{L}$ which justifies (B.7) for g_z was chosen to be an element of $\mathrm{Dom}(\dot{A})$.

Denote by B the restriction of $\dot{A}$ on

$$\mathrm{Dom}(B) = \mathcal{L} \cap \mathrm{Dom}(\dot{A}).$$

Our next claim is that $\mathrm{Dom}(B)$ is dense in $\mathcal{L}$.

Indeed, let $f \in \mathcal{L}$ and $f \perp \mathrm{Dom}(B)$, that is,

$$(f, g) = 0 \quad \text{for all } g \in \mathrm{Dom}(B).$$

From (B.7) follows that $f \in \mathrm{Ran}(B - iI)$ and hence $f = (B - iI)h$ for some $h \in \mathrm{Dom}(B)$. Thus, for all $g \in \mathrm{Dom}(B)$ one obtains that

$$(f, g) = ((B - iI)h, g) = (h, (B + iI)g) = 0. \tag{B.12}$$

On the other hand, (B.7) yields

$$\mathrm{Ran}(B + iI) = \mathcal{L}$$

and therefore from (B.12) follows that $h = 0$ and hence $f = (B - iI)h = 0$.

So, we have shown that the operator B is a densely defined symmetric operator in the Hilbert space $\mathcal{L}$ such that $\mathrm{Ran}(B \pm iI) = \mathcal{L}$ which means that B is self-adjoint.

Now suppose that a subspace $\mathcal{K}_0$ reduces $\dot{A}$ and that the part $\dot{A}|_{\mathcal{K}_0}$ is self-adjoint. Then

$$\mathrm{Ran}(\dot{A}|_{\mathcal{K}_0} - zI_{\mathcal{K}_0}) = \mathcal{K}_0, \quad z \in \mathbb{C} \setminus \mathbb{R},$$

which means that $\mathcal{K}_0 \subset \mathcal{L}$, proving that the part $\dot{A}$ in the subspace $\mathcal{K}$ is a prime symmetric operator.

The proof is complete. $\qquad\square$

Remark B.3. In the situation of Theorem B.2 the operator $\dot{A}|_{\mathcal{K}}$ is called the *prime part* of the symmetric operator $\dot{A}$.

Recall that an element $h \in \mathcal{H}$ is said to be a generating element for a self-adjoint operator H in the Hilbert space $\mathcal{H}$ if

$$\overline{\mathrm{span}_{\mathrm{Im}(z)\neq0}\{(H-zI)^{-1}h\}} = \mathcal{H}.$$

We also say that a self-adjoint operator has *simple spectrum* if it has a generating element.

Corollary B.4 (cf. [135]). *Let $\dot{A}$ be a closed symmetric operator with equal deficiency indices in a Hilbert space $\mathcal{H}$.*

Then $\dot{A}$ is a prime operator if and only if

$$\mathcal{H} = \overline{\mathrm{span}_{\mathrm{Im}(z)\neq0}\,\mathrm{Ker}((\dot{A})^* - zI)}.$$

If, in addition, $\dot{A}$ has deficiency indices $(1,1)$, then $\dot{A}$ is a prime operator if and only if for any self-adjoint extension A of $\dot{A}$ a deficiency element $0 \neq g_+ \in \mathrm{Ker}((\dot{A})^* - iI)$ is generating, that is,

$$\mathcal{H} = \overline{\mathrm{span}_{\mathrm{Im}(z)\neq0}(A-zI)^{-1}g_+}. \tag{B.13}$$

In particular, in this case, any self-adjoint extension of $\dot{A}$ has simple spectrum.

Proof. The first assertion has already been proven in Theorem B.2.

To prove the remaining statement, one proceeds as follows.

Suppose that (B.13) fails to hold and therefore the orthogonality condition

$$((A-zI)^{-1}g_+, h) = 0 \quad \text{for all } z \in \mathbb{C}\setminus\mathbb{R} \tag{B.14}$$

holds for some element $h \in \mathcal{H}$, $h \neq 0$. Since

$$\begin{aligned}
((A-zI)^{-1}g_+, h) &= ((A-zI)^{-1}g_+, (A+iI)(A+iI)^{-1}h) \\
&= ((A-iI)(A-zI)^{-1}g_+, (A+iI)^{-1}h),
\end{aligned}$$

one concludes that

$$((A-iI)(A-zI)^{-1}g_+, g) = 0 \quad \text{for all } z \in \mathbb{C}\setminus\mathbb{R}, \tag{B.15}$$

where

$$g = (A+iI)^{-1}h.$$

Observing that

$$(A-iI)(A-zI)^{-1}g_+ \in \mathrm{Ker}((\dot{A})^* - zI),$$

the hypothesis that $\dot{A}$ has deficiency indices $(1,1)$ yields the orthogonality condition

$$g \perp \overline{\mathrm{span}_{\mathrm{Im}(z)\neq 0} \mathrm{Ker}((\dot{A})^* - zI)},$$

and therefore $\dot{A}$ is not a prime operator by Theorem B.2.

Conversely, suppose that $\dot{A}$ is not a prime operator and therefore (B.15) holds for some $0 \neq g \in \mathcal{H}$. In particular,

$$(g_+, g) = 0. \tag{B.16}$$

Using the first resolvent identity,

$$(A - zI)^{-1} - (A - iI)^{-1} = (z - i)(A - iI)^{-1}(A - zI)^{-1},$$

one obtains

$$(A - iI)(A - zI)^{-1} = I + (z - i)(A - zI)^{-1}.$$

Therefore,

$$((A - iI)(A - zI)^{-1}g_+, g) = (g_+, g) + ((z - i)(A - zI)^{-1}g_+, g)$$

for all $z \in \mathbb{C} \setminus \mathbb{R}$. Using (B.15) and (B.16), this equality yields

$$((A - zI)^{-1}g_+, g) = 0 \quad \text{for all } z \in \mathbb{C} \setminus \mathbb{R}, \quad z \neq i,$$

and therefore, by continuity,

$$((A - zI)^{-1}g_+, g) = 0 \quad \text{for all } z \in \mathbb{C} \setminus \mathbb{R},$$

which shows that

$$\mathcal{H} \neq \overline{\mathrm{span}_{\mathrm{Im}(z)\neq 0}(A - zI)^{-1}g_+}.$$

So, we have shown that $\dot{A}$ is a prime operator if and only if (B.13) holds. The proof is complete. $\qquad\square$

We will also need a variant of the first part of this corollary in case when $\dot{A}$ has deficiency indices $(0,1)$ or $(1,0)$.

Lemma B.5. *Let $\dot{A}$ be a symmetric operator with deficiency indices $(0,1)$ in a Hilbert space $\mathcal{H}$. Then $\dot{A}$ is a prime operator if and only if*

$$\mathcal{H} = \overline{\mathrm{span}_{\mathrm{Im}(z)<0} \mathrm{Ker}((\dot{A})^* - zI)}. \tag{B.17}$$

Proof. *"Only if"* Part. Without loss one may assume that $\dot{A}$ is a closed operator. It is well known (see, [3, Theorem 2, Ch. VIII, Sec. 104]) that $\dot{A}$ is unitarily equivalent to the differentiation operator on the positive semi-axis with the Dirichlet boundary condition at the origin. So, without loss of generality, one may assume that $\mathcal{H} = L^2(\mathbb{R}_+)$ and

$$(\dot{A}f)(x) = -\frac{1}{i}\frac{d}{dx}f(x) \quad \text{a. e. } x \in \mathbb{R}_+ \tag{B.18}$$

on

$$\text{Dom}(\dot{A}) = \{f \in W_2^1(\mathbb{R}_+), \ f(0) = 0\}.$$

Clearly, the functions

$$h_z(x) = e^{-izx}, \quad x \in (0, \infty),$$

generate the subspaces $\text{Ker}((\dot{A})^* - zI)$, $\text{Im}(z) < 0$.

Now, if $f \in L^2((0, \infty))$ is orthogonal to h_z for all $z \in \mathbb{C}_-$, then

$$\int_0^\infty e^{-izx}\overline{f(x)}dx = 0 \quad \text{for all } z \in \mathbb{C}_-.$$

In particular,

$$H(s) = \int_0^\infty e^{-sx}\overline{f(x)}dx = 0, \quad s > 0,$$

and hence $f = 0$ by the uniqueness theorem for the Laplace transform (see, e.g., [22, Th. 5.5]). Therefore, (B.17) holds which completes the proof.

"If" Part. Suppose that $\dot{A}$ is not a prime operator. Therefore, by Theorem B.2,

$$\mathcal{H} \neq \overline{\text{span}_{\text{Im}(z)\neq 0}\text{Ker}((\dot{A})^* - zI)}.$$

Since $\dot{A}$ has deficiency indices $(0, 1)$,

$$\overline{\text{span}_{\text{Im}(z)\neq 0}\text{Ker}((\dot{A})^* - zI)} = \overline{\text{span}_{\text{Im}(z)<0}\text{Ker}((\dot{A})^* - zI)},$$

and hence

$$\mathcal{H} \neq \overline{\text{span}_{\text{Im}(z)<0}\text{Ker}((\dot{A})^* - zI)}.$$

Therefore, (B.17) fails to hold, which completes the proof of the *"If"* Part. □

In a similar way one proves the following statement.

Lemma B.6. *Let $\dot{A}$ be a symmetric operator with deficiency indices $(1,0)$ in a Hilbert space $\mathcal{H}$. Then $\dot{A}$ is a prime operator if and only if*

$$\mathcal{H} = \overline{\mathrm{span}_{\mathrm{Im}(z)>0} \, \mathrm{Ker}((\dot{A})^* - zI)}. \tag{B.19}$$

Appendix C

A FUNCTIONAL MODEL
OF A TRIPLE

Recall the notion of a *functional model* of a prime dissipative triple parameterized by the characteristic function [86].

Given a contractive analytic map $S(z)$,

$$S(z) = \frac{s(z) - \varkappa}{\overline{\varkappa}\, s(z) - 1}, \quad z \in \mathbb{C}_+, \tag{C.1}$$

where $|\varkappa| < 1$ and $s(z)$ is an analytic, contractive function in $\mathbb{C}_+$ satisfying the Livšic criterion [73] (also see [86, Theorem 1.2]), that is,

$$s(i) = 0 \quad \text{and} \quad \lim_{z \to \infty} z(s(z) - e^{2i\alpha}) = \infty \quad \text{for all } \alpha \in [0, \pi),$$

$$0 < \varepsilon \leq \arg(z) \leq \pi - \varepsilon, \tag{C.2}$$

introduce the function

$$M(z) = \frac{1}{i} \cdot \frac{s(z) + 1}{s(z) - 1}, \quad z \in \mathbb{C}_+. \tag{C.3}$$

One observes that

$$M(z) = \int_{\mathbb{R}} \left(\frac{1}{\lambda - z} - \frac{\lambda}{1 + \lambda^2} \right) d\mu(\lambda), \quad z \in \mathbb{C}_+,$$

for some infinite Borel measure,

$$\mu(\mathbb{R}) = \infty, \tag{C.4}$$

such that

$$\int_{\mathbb{R}} \frac{d\mu(\lambda)}{1 + \lambda^2} = 1. \tag{C.5}$$

In the Hilbert space $L^2(\mathbb{R}; d\mu)$ introduce the (self-adjoint) operator $\mathcal{B}$ of multiplication by independent variable on

$$\text{Dom}(\mathcal{B}) = \left\{ f \in L^2(\mathbb{R}; d\mu) \Big| \int_{\mathbb{R}} \lambda^2 |f(\lambda)|^2 d\mu(\lambda) < \infty \right\}. \tag{C.6}$$

Denote by $\dot{\mathcal{B}}$ its restriction on

$$\text{Dom}(\dot{\mathcal{B}}) = \left\{ f \in \text{Dom}(\mathcal{B}) \Big| \int_{\mathbb{R}} f(\lambda) d\mu(\lambda) = 0 \right\}, \tag{C.7}$$

and let $\widehat{\mathcal{B}}$ be the dissipative quasi-selfadjoint extension of the symmetric operator $\dot{\mathcal{B}}$ on

$$\text{Dom}(\widehat{\mathcal{B}}) = \text{Dom}(\dot{\mathcal{B}}) \dotplus \text{lin span} \left\{ \frac{1}{\lambda - i} - \varkappa \frac{1}{\lambda + i} \right\}, \tag{C.8}$$

where the von Neumann parameter $\varkappa$ of the triple $(\dot{\mathcal{B}}, \widehat{\mathcal{B}}, \mathcal{B})$ is given by

$$\varkappa = S(i).$$

Notice that in this case

$$\text{Dom}(\mathcal{B}) = \text{Dom}(\dot{\mathcal{B}}) \dotplus \text{lin span} \left\{ \frac{1}{\lambda - i} - \frac{1}{\lambda + i} \right\}. \tag{C.9}$$

We will refer to the triple $(\dot{\mathcal{B}}, \widehat{\mathcal{B}}, \mathcal{B})$ as *the model triple* in the Hilbert space $L^2(\mathbb{R}; d\mu)$.

Let $\dot{A}$ be a densely defined symmetric operator with deficiency indices $(1,1)$, A its self-adjoint (reference) extension and $\widehat{A}$ a maximal non-selfadjoint dissipative extension of $\dot{A}$.

Denote by $S_{\mathfrak{A}}(z) = S_{(\dot{A}, \widehat{A}, A)}(z)$ the characteristic function of the triple $\mathfrak{A} = (\dot{A}, \widehat{A}, A)$ as introduced in Chapter 3 by eq. (2.9). Notice that the characteristic function $S_{\mathfrak{B}}(z)$ of the model triple $\mathfrak{M} = (\dot{\mathcal{B}}, \widehat{\mathcal{B}}, \mathcal{B})$ is given by (C.1).

The following uniqueness result shows that the characteristic function of a triple $(\dot{A}, \widehat{A}, A)$ is a complete unitary invariant of the triple whenever the symmetric operator $\dot{A}$ is prime, equivalently, the dissipative operator $\widehat{A}$ is completely non-selfadjoint.

Theorem C.1 ([86, Theorems 1.4, 4.1]). *Suppose that $\dot{A}$ and $\dot{B}$ are prime, closed, densely defined symmetric operators with deficiency indices $(1,1)$. Assume, in addition, that A and B are some self-adjoint extensions of $\dot{A}$ and $\dot{B}$ and that $\widehat{A}$ and $\widehat{B}$ are maximal dissipative extensions of $\dot{A}$ and $\dot{B}$, respectively $(\widehat{A} \neq (\widehat{A})^*, \ \widehat{B} \neq (\widehat{B})^*)$.*

Then

(i) *the triples* $\mathfrak{A} = (\dot{A}, \widehat{A}, A)$ *and* $\mathfrak{B} = (\dot{B}, \widehat{B}, B)$ *are mutually unitarily equivalent[1] if, and only if, the characteristic functions* $S_{\mathfrak{A}}(z) = S_{(\dot{A}, \widehat{A}, A)}(z)$ *and* $S_{\mathfrak{B}}(z) = S_{(\dot{B}, \widehat{B}, B)}(z)$ *of the triples coincide;*

(ii) *the triple* $(\dot{A}, \widehat{A}, A)$ *is mutually unitarily equivalent to the model triple*

$$\mathfrak{M} = (\dot{\mathcal{B}}, \widehat{\mathcal{B}}, \mathcal{B})$$

in the Hilbert space $L^2(\mathbb{R}; d\mu)$, *where* $\mu(d\lambda)$ *is the representing measure for the Weyl-Titchmarsh function* $M(z) = M_{(\dot{A}, A)}(z)$ *associated with the pair* $(\dot{A}, A)$.

In particular,

(iii) *the pairs* $(\dot{A}, A)$ *and* $(\dot{B}, B)$ *are mutually unitary equivalent if and only if* $M_{(\dot{A}, A)}(z) = M_{(\dot{B}, B)}(z)$.

Remark C.2. Notice that in view of (C.9), if $\mathcal{U}$ is the unitary operator from the Hilbert space $\mathcal{H}$ onto $L^2(\mathbb{R}; d\mu)$ that implements mutual unitary equivalence of the triples $(\dot{A}, \widehat{A}, A)$ and $(\dot{B}, \widehat{B}, B)$, then

$$(\mathcal{U}g_\pm)(\lambda) = \frac{\Theta}{\lambda \mp i} \quad \text{for some } |\Theta| = 1,$$

provided that $g_\pm \in \mathrm{Ker}((\dot{A})^* \mp iI)$ are normalized deficiency elements of $\dot{A}$ such that

$$g_+ - g_- \in \mathrm{Dom}(A). \tag{C.10}$$

Indeed, for the normalized deficiency elements $g_\pm$ and $\mathcal{U}g_\pm$ of the symmetric operators $\dot{A}$ and $\dot{B}$, respectively, we have

$$(\mathcal{U}g_\pm)(\lambda) = \frac{\Theta_\pm}{\lambda \mp i} \quad \text{for some } |\Theta_\pm| = 1.$$

Since $\mathcal{B} = \mathcal{U}A\mathcal{U}^{-1}$, from (C.10) it follows that the function $h = \mathcal{U}(g_+ - g_-)$ given by

$$h(\lambda) = \frac{\Theta_+}{\lambda - i} - \frac{\Theta_-}{\lambda + i}, \quad \lambda \in \mathbb{R},$$

---

[1] We say that triples of operators $(\dot{A}, \widehat{A}, A)$ and $(\dot{B}, \widehat{B}, B)$ in Hilbert spaces $\mathcal{H}_A$ and $\mathcal{H}_B$ are mutually unitarily equivalent if there is a unitary map $\mathcal{U}$ from $\mathcal{H}_A$ onto $\mathcal{H}_B$ such that $\dot{B} = \mathcal{U}\dot{A}\mathcal{U}^{-1}$, $\widehat{B} = \mathcal{U}\widehat{A}\mathcal{U}^{-1}$, and $B = \mathcal{U}A\mathcal{U}^{-1}$.

belongs to $\mathrm{Dom}(\mathcal{B}) = L^2(\mathbb{R}; (1+\lambda^2)d\mu(\lambda))$ and therefore

$$\int_{\mathbb{R}} \left| \frac{\Theta_+}{\lambda - i} - \frac{\Theta_-}{\lambda + i} \right|^2 (1+\lambda^2)d\mu(\lambda) < \infty.$$

Taking into account that by (C.4) the measure $\mu(d\lambda)$ is infinite, one necessarily gets that $\Theta_+ = \Theta_-$ and the claim follows.

Appendix D

THE SPECTRAL ANALYSIS OF THE MODEL DISSIPATIVE OPERATOR

In the suggested functional model for a triple in $L^2(\mathbb{R}; d\mu)$, the eigenfunctions of the model dissipative operator $\widehat{\mathcal{B}}$ from the model triple $(\dot{\mathcal{B}}, \widehat{\mathcal{B}}, \mathcal{B})$ given by (C.6)–(C.8) look exceptionally simple [86].

Lemma D.1. *Suppose that* $\mathfrak{M} = (\dot{\mathcal{B}}, \widehat{\mathcal{B}}, \mathcal{B})$ *is the model triple in* $L^2(\mathbb{R}; d\mu)$ *given by* (C.6)–(C.8). *Then a point* $z_0 \in \mathbb{C}_+$ *is an eigenvalue of the dissipative operator* $\widehat{\mathcal{B}}$ *if and only if*

$$S_{\mathfrak{M}}(z_0) = S_{(\dot{\mathcal{B}}, \widehat{\mathcal{B}}, \mathcal{B})}(z_0) = 0.$$

In this case, the corresponding eigenfunction f *is of the form*

$$f(\lambda) = \frac{1}{\lambda - z_0} \quad \text{for } \mu\text{-almost all } \lambda \in \mathbb{R}.$$

Proof. Suppose that $z_0 \in \mathbb{C}_+$ is an eigenvalue of $\widehat{\mathcal{B}}$ and $f \in L^2(\mathbb{R}; d\mu)$ is the corresponding eigenvector,

$$\widehat{\mathcal{B}}f = z_0 f, \quad f \in \text{Dom}(\widehat{\mathcal{B}}).$$

Since $f \in \text{Dom}(\widehat{\mathcal{B}})$, the element f admits the representation

$$f(\lambda) = f_0(\lambda) + K \left(\frac{1}{\lambda - i} - \varkappa \frac{1}{\lambda + i} \right),$$

where $f_0 \in \text{Dom}(\dot{\mathcal{B}})$, K is some constant and $\varkappa = S_{\mathcal{B}}(i)$. Then

$$0 = ((\widehat{\mathcal{B}} - z_0 I)f)(\lambda) = (\lambda - z_0)f_0(\lambda) + K \left(\frac{i - z_0}{\lambda - i} + \varkappa \frac{i + z_0}{\lambda + i} \right).$$

Hence,

$$f_0(\lambda) = -\frac{K}{\lambda - z_0}\left(\frac{i - z_0}{\lambda - i} + \varkappa\frac{i + z_0}{\lambda + i}\right).$$

Since $f_0 \in \mathrm{Dom}(\dot{\mathcal{B}})$, one obtains that

$$\int_{\mathbb{R}} f_0(\lambda)d\mu(\lambda) = 0$$

and hence

$$\begin{aligned}
0 &= \int_{\mathbb{R}} \frac{1}{\lambda - z_0}\left(\frac{i - z_0}{\lambda - i} + \varkappa\frac{i + z_0}{\lambda + i}\right)d\mu(\lambda) \\
&= -\int_{\mathbb{R}}\left(\frac{1}{\lambda - z_0} - \frac{1}{\lambda - i}\right)d\mu(\lambda) + \varkappa\int_{\mathbb{R}}\left(\frac{1}{\lambda - z_0} - \frac{1}{\lambda + i}\right)d\mu(\lambda) \\
&= -M(z_0) + M(i) + \varkappa(M(z_0) - M(-i)) \\
&= -M(z_0) + i + \varkappa(M(z_0) + i),
\end{aligned}$$

where $M(z) = M_{(\dot{\mathcal{B}},\mathcal{B})}(z)$ is the Weyl-Titchmarsh function associated with the model pair $(\dot{\mathcal{B}}, \mathcal{B})$.

Therefore,

$$\varkappa = \frac{M(z_0) - i}{M(z_0) + i} = s_{(\dot{\mathcal{B}},\mathcal{B})}(z_0),$$

where $s_{(\dot{\mathcal{B}},\mathcal{B})}(z)$ is the Livšic function associated with the pair $(\dot{\mathcal{B}}, \mathcal{B})$. Hence, the characteristic function $S_{\mathcal{B}}(z)$ vanishes at the point z_0,

$$S_{\mathcal{B}}(z_0) = \frac{s_{(\dot{\mathcal{B}},\mathcal{B})}(z_0) - \varkappa}{\overline{\varkappa}s_{(\dot{\mathcal{B}},\mathcal{B})}(z_0) - 1} = 0.$$

In this case,

$$\begin{aligned}
f(\lambda) &= K\left[\left(\frac{1}{\lambda - i} - \varkappa\frac{1}{\lambda + i}\right) - \frac{1}{\lambda - z_0}\left(\frac{i - z_0}{\lambda - i} + \varkappa\frac{i + z_0}{\lambda + i}\right)\right] \\
&= K\left[\frac{1}{\lambda - i}\left(1 - \frac{i - z_0}{\lambda - z_0}\right) - \varkappa\frac{1}{\lambda + i}\left(1 + \frac{i + z_0}{\lambda - z_0}\right)\right] \\
&= K(1 - \varkappa)\cdot\frac{1}{\lambda - z_0}.
\end{aligned}$$

So, we have shown that if z_0 is an eigenvalue of $\widehat{\mathcal{B}}$, then

$$S_{\mathfrak{M}}(z_0) = 0,$$

with the corresponding eigenelement f of the form

$$f(\lambda) = \frac{1}{\lambda - z_0}. \tag{D.1}$$

Repeating the same reasoning in the reversed order, one shows that if $S_{\mathfrak{M}}(z_0) = 0$, then the function f given by (D.1) belongs to $\mathrm{Dom}(\widehat{\mathcal{B}})$ and $\widehat{\mathcal{B}} f = z_0 f$. $\qquad\qquad\square$

For the resolvents of the model dissipative operator $\widehat{\mathcal{B}}$ and the self-adjoint (reference) operator $\mathcal{B}$ from the model triple $\mathfrak{M} = (\dot{\mathcal{B}}, \widehat{\mathcal{B}}, \mathcal{B})$ one gets the following resolvent formula [86].

Theorem D.2. *Suppose that* $\mathfrak{M} = (\dot{\mathcal{B}}, \widehat{\mathcal{B}}, \mathcal{B})$ *is the model triple in the Hilbert space* $L^2(\mathbb{R}; d\mu)$ *given by* (C.6)–(C.8).

Then the resolvent of the model dissipative operator $\widehat{\mathcal{B}}$ *in* $L^2(\mathbb{R}; d\mu)$ *has the form*

$$(\widehat{\mathcal{B}} - zI)^{-1} = (\mathcal{B} - zI)^{-1} - p(z)(\cdot, g_{\bar{z}})g_z, \tag{D.2}$$

with

$$p(z) = \left(M_{(\dot{\mathcal{B}}, \mathcal{B})}(z) + i\frac{\varkappa + 1}{\varkappa - 1} \right)^{-1}, \tag{D.3}$$

$$z \in \rho(\widehat{\mathcal{B}}) \cap \rho(\mathcal{B}).$$

Here $M_{(\dot{\mathcal{B}}, \mathcal{B})}(z)$ *is the Weyl-Titchmarsh function associated with the pair* $(\dot{\mathcal{B}}, \mathcal{B})$ *continued to the lower half-plane by the Schwarz reflection principle,* $\varkappa$ *is the von Neumann parameter of the triple* $\mathfrak{M}$, *and the deficiency elements* g_z,

$$g_z \in \mathrm{Ker}((\dot{\mathcal{B}})^* - zI), \quad z \in \mathbb{C} \setminus \mathbb{R},$$

are given by

$$g_z(\lambda) = \frac{1}{\lambda - z} \quad \text{for } \mu\text{-almost all } \lambda \in \mathbb{R}. \tag{D.4}$$

Proof. Given $h \in L^2(\mathbb{R}; d\mu)$ and $z \in \rho(\widehat{\mathcal{B}})$, suppose that

$$(\widehat{\mathcal{B}} - zI)f = h \quad \text{for some } f \in \mathrm{Dom}(\widehat{\mathcal{B}}). \tag{D.5}$$

Since $f \in \mathrm{Dom}(\widehat{\mathcal{B}})$, one gets the representation

$$f(\lambda) = f_0(\lambda) + K \left(\frac{1}{\lambda - i} - \varkappa \frac{1}{\lambda + i} \right) \qquad \text{(D.6)}$$

for some $f_0 \in \mathrm{Dom}(\dot{\mathcal{B}})$ and $K \in \mathbb{C}$. Eq. (D.5) yields

$$(\lambda - z)f_0(\lambda) + K \left(\frac{i - z}{\lambda - i} + \varkappa \frac{i + z}{\lambda + i} \right) = h(\lambda)$$

and hence

$$f_0(\lambda) = \frac{h(\lambda)}{\lambda - z} - \frac{K}{\lambda - z} \left(\frac{i - z}{\lambda - i} + \varkappa \frac{i + z}{\lambda + i} \right). \qquad \text{(D.7)}$$

Since $f_0 \in \mathrm{Dom}(\dot{\mathcal{B}})$,

$$\int_{\mathbb{R}} f_0(\lambda) d\mu(\lambda) = 0.$$

Integrating (D.7) against the measure $\mu(d\lambda)$, one obtains that

$$K \int_{\mathbb{R}} \frac{1}{\lambda - z} \left(\frac{i - z}{\lambda - i} + \varkappa \frac{i + z}{\lambda + i} \right) d\mu(\lambda) = \int_{\mathbb{R}} \frac{h(\lambda)}{\lambda - z} d\mu(\lambda). \qquad \text{(D.8)}$$

Observing that

$$\int_{\mathbb{R}} \frac{1}{\lambda - z} \left(\frac{i - z}{\lambda - i} + \varkappa \frac{i + z}{\lambda + i} \right) d\mu(\lambda) = i - M(z) + \varkappa(M(z) + i),$$

with $M(z) = M_{(\dot{\mathcal{B}}, \widehat{\mathcal{B}})}(z)$, and solving (D.8) for K, one obtains

$$K = \frac{1}{(\varkappa - 1)M(z) + i(1 + \varkappa)} \int_{\mathbb{R}} \frac{h(\lambda)}{\lambda - z} d\mu(\lambda).$$

Combining (D.6) and (D.7), for the element f we have the representation

$$
\begin{aligned}
f(\lambda) &= \frac{h(\lambda)}{\lambda - z} + \frac{K}{\lambda - z} \left(\frac{\lambda - z}{\lambda - i} - \varkappa \frac{\lambda - z}{\lambda + i} - \left[\frac{i - z}{\lambda - i} + \varkappa \frac{i + z}{\lambda + i} \right] \right) \\
&= \frac{h(\lambda)}{\lambda - z} - K \frac{\varkappa - 1}{\lambda - z} \\
&= \frac{h(\lambda)}{\lambda - z} - \left(M(z) + i \frac{\varkappa + 1}{\varkappa - 1} \right)^{-1} \frac{1}{\lambda - z} \int_{\mathbb{R}} \frac{h(\lambda)}{\lambda - z} d\mu(\lambda), \\
&\qquad z \in \rho(\widehat{\mathcal{B}}) \cap \rho(\mathcal{B}), \qquad \text{(D.9)}
\end{aligned}
$$

where we have used (D.8) on the last step.

To complete the proof, it remains to recall that $f = (\widehat{\mathcal{B}} - zI)^{-1}h$ and to compare (D.2) with (D.9). □

Remark D.3. Using (C.3), it is easy to see that the poles of the function $p(z)$, defined in (D.3), in the upper half-plane coincide with the roots of the equation

$$s_{(\dot{\mathcal{B}},\mathcal{B})}(z) = \varkappa, \quad z \in \mathbb{C}_+,$$

provided that $\varkappa \neq 0$ and $M(z) \neq i$ identically in the upper half-plane. Therefore, the zeros of the characteristic function $S_{\mathfrak{M}}(z) = S_{(\mathcal{B},\widehat{\mathcal{B}},\mathcal{B})}(z)$ in the upper half-pane determine the poles of the resolvent of the dissipative operator $\widehat{\mathcal{B}}$ (cf. Lemma D.1).

We also remark that if $\varkappa = 0$ and $M(z) = i$ for all $z \in \mathbb{C}_+$, then the point spectrum of the dissipative operator $\widehat{\mathcal{B}}$ fills in the whole open upper half-plane $\mathbb{C}_+$.

Given a triple $(\dot{A}, \widehat{A}, A)$ satisfying (2.1) and (2.5), the following corollary provides an analog of the Krein formula for resolvents for all quasi-selfadjoint dissipative extensions of the symmetric operator $\dot{A}$ with deficiency indices $(1, 1)$.

Corollary D.4 ([86]). *Let $\mathfrak{A} = (\dot{A}, \widehat{A}, A)$ be a triple satisfying (2.1) and (2.5). Then the following resolvent formula*

$$(\widehat{A} - zI)^{-1} = (A - zI)^{-1} - p(z)(\cdot, g_{\bar{z}})g_z,$$

$$z \in \rho(\widehat{A}) \cap \rho(A), \tag{D.10}$$

holds.
   Here

(a) *the function $p(z)$ is given by*

$$p(z) = \left(M_{(\dot{A},A)}(z) + i\frac{\varkappa + 1}{\varkappa - 1} \right)^{-1} \tag{D.11}$$

$$= i \left(\frac{s_{(\dot{A},A)}(z) + 1}{s_{(\dot{A},A)}(z) - 1} - \frac{\varkappa + 1}{\varkappa - 1} \right)^{-1}; \tag{D.12}$$

(b) *$M_{(\dot{A},A)}(z)$ and $s_{(\dot{A},A)}(z)$ are the Weyl-Titchmarsh and the Livšic function of the pair $(\dot{A}, A)$, respectively;*

(c) g_z are deficiency elements of $\dot{A}$,

$$g_z \in \mathrm{Ker}((\dot{A})^* - zI),$$

satisfying the normalization condition

$$\|g_z\| = \left(\int_{\mathbb{R}} \frac{d\mu(\lambda)}{|\lambda - z|^2} \right)^{1/2} \tag{D.13}$$

(the deficiency elements g_z can be chosen to be analytic in $z \in \rho(\widehat{A}) \cap \rho(A)$);

(d) $\mu(d\lambda)$ is the measure from the Herglotz-Nevanlinna representation

$$M_{(\dot{A},A)}(z) = \int_{\mathbb{R}} \left(\frac{1}{\lambda - z} - \frac{\lambda}{1 + \lambda^2} \right) d\mu(\lambda);$$

(e) $\varkappa$ is the von Neumann parameter of the triple $(\dot{A}, \widehat{A}, A)$ which characterizes the domain of the dissipative extension $\widehat{A}$ in such a way that

$$g_+ - g_- \in \mathrm{Dom}(A) \quad \text{and} \quad g_+ - \varkappa g_- \in \mathrm{Dom}(\widehat{A}), \tag{D.14}$$

where $g_\pm = g_{\pm i}$.

Remark D.5. We would like to stress that the von Neumann parameter $\varkappa$ of the triple $\mathfrak{A} = (\dot{A}, \widehat{A}, A)$, the Livšic function $s_{(\dot{A},A)}(z)$, and the Weyl-Titchmarsh function $M_{(\dot{A},A)}(z)$, can easily be recovered from the knowledge of the the characteristic function $S(z) = S_{\mathfrak{A}}(z)$. (Recall that the characteristic function $S(z)$ is a complete unitary invariant of the triple $\mathfrak{A} = (\dot{A}, \widehat{A}, A)$, provided that $\dot{A}$ is a prime operator).
Indeed,

$$\varkappa = S(i),$$

$$s_{(\dot{A},A)}(z) = \frac{S(z) - \varkappa}{\overline{\varkappa} S(z) - 1},$$

$$M_{(\dot{A},A)}(z) = \frac{1}{i} \cdot \frac{s_{(\dot{A},A)}(z) + 1}{s_{(\dot{A},A)}(z) - 1},$$

$$z \in \mathbb{C}_+,$$

with $M_{(\dot{A},A)}(z)$ continued to the lower half-plane by the Schwarz reflection principle

$$M_{(\dot{A},A)}(z) = \overline{M_{(\dot{A},A)}(\overline{z})}, \quad z \in \mathbb{C}_-.$$

Remark D.6. The resolvent formula (D.10)–(D.11) also holds if $|\varkappa| = 1$ and hence $\widehat{A}$ is self-adjoint. In this case, it coincides with the Krein resolvent formula for self-adjoint extensions of $\dot{A}$.

Remark D.7. If two triples $(\dot{A}, \widehat{A}_1, A)$ and $(\dot{A}, \widehat{A}_2, A)$ with the same reference operator A satisfy (2.1) and (2.5) and have the von Neumann parameters $\varkappa_1$ and $\varkappa_2$, respectively, then one gets the following resolvent formula for the dissipative extensions $\widehat{A}_1$ and $\widehat{A}_2$ refining, in the rank-one setting, a result in [62]:

$$(\widehat{A}_2 - zI)^{-1} = (\widehat{A}_1 - zI)^{-1} - q(z)(\cdot, g_{\bar{z}})g_z,$$

where $q(z) = p_2(z) - p_1(z)$ with

$$p_k(z) = \left(M_{(\dot{A},A)}(z) + i\frac{\varkappa_k + 1}{\varkappa_k - 1} \right)^{-1},$$

$$= i\left(\frac{s_{(\dot{A},A)}(z) + 1}{s_{(\dot{A},A)}(z) - 1} - \frac{\varkappa_k + 1}{\varkappa_k - 1} \right)^{-1}, \quad k = 1, 2,$$

$$z \in \rho(A) \cap \rho(\widehat{A}_1) \cap \rho(\widehat{A}_2).$$

We recall, see (2.9), that if $S_1(z)$ and $S_2(z)$ are the characteristic functions of the triples $(\dot{A}, \widehat{A}_1, A)$ and $(\dot{A}, \widehat{A}_2, A)$, respectively, then

$$s_{(\dot{A},A)}(z) = \frac{S_k(z) - \varkappa_k}{\overline{\varkappa_k}S_k(z) - 1}, \quad k = 1, 2.$$

Corollary D.8. *Suppose that* $\mathfrak{M} = (\dot{\mathcal{B}}, \widehat{\mathcal{B}}, \mathcal{B})$ *is the model triple in* $L^2(\mathbb{R}; d\mu)$ *given by* (C.6)–(C.8). *Assume that* $z = 0$ *is a regular point for both the dissipative operator* $\widehat{\mathcal{B}}$ *and the (reference) self-adjoint operator* $\mathcal{B}$. *Then the inverse* $\widehat{\mathcal{B}}^{-1}$ *is a rank-one perturbation of the (bounded) self-adjoint operator* $\mathcal{B}^{-1}$. *That is,*

$$\widehat{\mathcal{B}}^{-1} = \mathcal{B}^{-1} - pQ,$$

where

$$p = \left(M(0) + i\frac{\varkappa + 1}{\varkappa - 1} \right)^{-1}, \tag{D.15}$$

Q *is a rank-one self-adjoint operator*

$$(Qf)(\lambda) = \frac{1}{\lambda} \int_{\mathbb{R}} \frac{f(s)}{s} d\mu(s), \quad \mu\text{-}a.e. \ \lambda \in \mathbb{R},$$

and $M(0)$ *is the value of the Weyl-Titchmarsh function associated with the pair* $(\dot{\mathcal{B}}, \mathcal{B})$ *at the point zero.*

Appendix E

TRANSFORMATION LAWS

In this chapter we discuss the dependence of the Livšic, Weyl-Titchmarsh and characteristic functions upon the reference operator.

Lemma E.1. *Assume that $\dot{A}$ is a symmetric, densely defined, closed operator with deficiency indices $(1,1)$ and $\widehat{A}$ its maximal dissipative extension such that $\widehat{A} \neq (\widehat{A})^*$. Suppose that $g_\pm \in \mathrm{Ker}((\dot{A})^* \mp iI)$, $\|g_\pm\| = 1$. Given $\alpha \in [0, \pi)$, denote by A_α a unique self-adjoint extension of $\dot{A}$ such that*

$$g_+ - e^{2i\alpha}g_- \in \mathrm{Dom}(A_\alpha). \tag{E.1}$$

Let $s_\alpha(z) = s_{(\dot{A}, A_\alpha)}(z)$, $M_\alpha(z) = M_{(\dot{A}, A_\alpha)}(z)$, and $S_\alpha(z) = S_{(\dot{A}, \widehat{A}, A_\alpha)}(z)$ be the Livšic function, the Weyl-Titchmarsh function associated with the pair $(\dot{A}, A_\alpha)$, and the characteristic function associated with the triple $(\dot{A}, \widehat{A}, A_\alpha)$, respectively.
   Then

$$s_\alpha(z) = e^{2i(\beta - \alpha)}s_\beta(z), \tag{E.2}$$

$$M_\beta(z) = \frac{\cos(\beta - \alpha)M_\alpha(z) - \sin(\beta - \alpha)}{\cos(\beta - \alpha) + \sin(\beta - \alpha)M_\alpha(z)}, \tag{E.3}$$

$$S_\alpha(z) = e^{2i(\beta - \alpha)}S_\beta(z). \tag{E.4}$$

Proof. Suppose that the deficiency elements $g_\pm \in \mathrm{Ker}((\dot{A})^* \mp iI)$, $\|g_\pm\| = 1$, are such that

$$g_+ - \varkappa g_- \in \mathrm{Dom}(\widehat{A})$$

for some $\varkappa$, $|\varkappa| < 1$. Denote by A_* the reference self-adjoint extension of $\dot{A}$ such that

$$g_+ - g_- \in \mathrm{Dom}(A_*),$$

and let $s_*(z) = s_{(\dot{A},A_*)}(z)$, $M_*(z) = M_{(\dot{A},A_*)}(z)$ and $S_*(z) = S_{(\dot{A},\widehat{A},A_*)}(z)$ be the corresponding Livšic, Weyl-Titchmarsh and the characteristic functions, respectively. From the definition of the Livšic function $s_\alpha(z)$ it follows that

$$s_\alpha(z) = e^{-2i\alpha} s_*(z).$$

By (2.4) we have

$$s_*(z) = \frac{M_*(z) - i}{M_*(z) + i} \quad \text{and} \quad s_\alpha(z) = e^{-2i\alpha} s_*(z) = \frac{M_\alpha(z) - i}{M_\alpha(z) + i}$$

and hence

$$M_\alpha(z) = \frac{\cos(\alpha) M_*(z) - \sin(\alpha)(z)}{\cos(\alpha) + \sin(\alpha) M_*(z)}.$$

From (2.9) it follows

$$S_*(z) = \frac{s_*(z) - \varkappa}{\overline{\varkappa} s_*(z) - 1}.$$

Therefore,

$$e^{-2i\alpha} S_*(z) = \frac{e^{-2i\alpha} s_*(z) - e^{-2i\alpha} \varkappa}{(e^{-2i\alpha} \varkappa) e^{-2i\alpha} s_*(z) - 1} = \frac{s_\alpha(z) - e^{-2i\alpha} \varkappa}{(e^{-2i\alpha} \varkappa) s_\alpha(z) - 1} = S_\alpha(z),$$

which proves (E.2), (E.3), (E.4) first for $\beta = 0$ and hence for all β taking into account that the transformations $\alpha \to s_\alpha, M_\alpha, S_\alpha$ are one-parameter groups. $\square$

Our next result shows that the concept of a characteristic function of a triple is essentially determined by the corresponding dissipative operator only rather than by the triple itself (cf. [3, 73]).

Proposition E.2. *Let $(\dot{A}, \widehat{A}, A)$ and $(\dot{B}, \widehat{B}, B)$ be two triples satisfying the hypothesis of Lemma E.1. Suppose that the dissipative operators $\widehat{A}$ and $\widehat{B}$ are unitarily equivalent.*

Then the characteristic functions of the triples $(\dot{A}, \widehat{A}, A)$ and $(\dot{B}, \widehat{B}, B)$ coincide up to a constant unimodular factor.

In particular, the absolute values of the von Neumann parameters $\varkappa_{(\dot{A},\widehat{A},A)}$ and $\varkappa_{(\dot{B},\widehat{B},B)}$ of the triples $(\dot{A}, \widehat{A}, A)$ and $(\dot{B}, \widehat{B}, B)$ coincide,

$$|\varkappa_{(\dot{A},\widehat{A},A)}| = |\varkappa_{(\dot{B},\widehat{B},B)}|. \tag{E.5}$$

Proof. To be more specific, assume that $\mathcal{U}$ is a unitary operator such that

$$\widehat{B} = \mathcal{U}^{-1}\widehat{A}\mathcal{U}.$$

That is,

$$\mathcal{U}(\mathrm{Dom}(\widehat{B})) = \mathrm{Dom}(\widehat{A})$$

and

$$\mathcal{U}\widehat{B}f = \widehat{A}\mathcal{U} \quad \text{for all } f \in \mathrm{Dom}(\widehat{B}).$$

Literally repeating the proof of Lemma 11.1 one shows that

$$(\widehat{B})^* = \mathcal{U}^{-1}(\widehat{A})^*\mathcal{U}.$$

Therefore, the symmetric operators $\dot{A}$ and $\dot{B}$ are unitarily equivalent

$$\dot{B} = \mathcal{U}^{-1}\dot{A}\mathcal{U},$$

since

$$\dot{A} = \widehat{A}\big|_{\mathrm{Dom}(\widehat{A})\cap\mathrm{Dom}((\widehat{A})^*)} \quad \text{and} \quad \dot{B} = \widehat{B}\big|_{\mathrm{Dom}(\widehat{B})\cap\mathrm{Dom}((\widehat{B})^*)}.$$

Moreover, the operator $B' = \mathcal{U}^{-1}A\mathcal{U}$ is a self-adjoint extensions of $\dot{B}$.

By Lemma E.1, the characteristic functions $S_{(\dot{B},\widehat{B},B)}(z)$ and $S_{(\dot{B},\widehat{B},B')}(z)$ of the triples $(\dot{B},\widehat{B},B)$ and $(\dot{B},\widehat{B},B')$ are related as

$$S_{(\dot{B},\widehat{B},B)}(z) = \Theta S_{(\dot{B},\widehat{B},B')}(z), \quad z \in \mathbb{C}_+,$$

for some constant Θ, $|\Theta| = 1$. Since the tripes $(\dot{B},\widehat{B},B')$ and $(\dot{A},\widehat{A},A)$ are mutually unitarily equivalent by construction, we have

$$S_{(\dot{B},\widehat{B},B')}(z) = S_{(\dot{A},\widehat{A},A)}(z), \quad z \in \mathbb{C}_+,$$

by the uniqueness Theorem C.1. Therefore,

$$S_{(\dot{B},\widehat{B},B)}(z) = \Theta S_{(\dot{A},\widehat{A},A)}(z), \quad z \in \mathbb{C}_+,$$

which completes the proof of the first assertion of the proposition.

To prove (E.5), we use the relation (2.8) to conclude that

$$|\varkappa_{(\dot{A},\widehat{A},A)}| = |S_{(\dot{A},\widehat{A},A)}(i)| = |S_{(\dot{B},\widehat{B},B)}(i)| = |\varkappa_{(\dot{B},\widehat{B},B)}|. \qquad \square$$

Our next goal is to establish a transformation law for the Livšic function under the affine transformations of the pair $(\dot{A}, A)$,

$$(\dot{A}, A) \longrightarrow (a\dot{A} + bI, aA + bI), \quad a, b \in \mathbb{R}, \quad a > 0.$$

Lemma E.3. *Let $\dot{A}$ be a symmetric operator with deficiency indices $(1,1)$ and A its self-adjoint extension. Suppose that $f(z) = az + b$ with $a, b \in \mathbb{R}$, $a > 0$ is an affine transformation.*

Then the Livšic function associated with the pair $(f(\dot{A}), f(A))$ admits the representation

$$s_{(f(\dot{A}), f(A))}(z) = \frac{m(z) - m(i)}{m(z) - \overline{m(i)}},\tag{E.6}$$

where

$$m(z) = M(f^{-1}(z))$$

and $M(z) = M_{(\dot{A}, A)}(z)$ is the Weyl-Titchmarsh function associated with the pair $(\dot{A}, A)$.

Proof. By Remark 2.5, without loss of generality one may assume that $\dot{A}$ is a prime symmetric operator. Let $M(z) = M_{(\dot{A}, A)}(z)$ be the Weyl-Titchmarsh function associated with the pair $(\dot{A}, A)$. Next, we may assume that A is the multiplication operator by independent variable in $L^2(\mathbb{R}, d\mu)$ and $\dot{A}$ is its restriction on

$$\mathrm{Dom}(\dot{A}) = \left\{ f \in \mathrm{Dom}(A) \,\middle|\, \int_{\mathbb{R}} f(\lambda) d\mu(\lambda) = 0 \right\},$$

where $\mu(d\lambda)$ is the representing measure for $M(z)$ (see Theorem C.1).

Introduce the family of functions

$$G_z(\lambda) = \frac{1}{\lambda - f^{-1}(z)}, \quad \mathrm{Im}(z) \neq 0.$$

Clearly,

$$G_z \in \mathrm{Ker}((f(\dot{A})^*) - zI), \quad \mathrm{Im}(z) \neq 0.$$

Set

$$G_+ = G_{f^{-1}(i)} \quad \text{and} \quad G_- = G_{f^{-1}(-i)}.$$

One easily checks that

$$\|G_+\| = \|G_-\|, \quad G_\pm \in \mathrm{Ker}(f(\dot{A}) \mp iI),$$

and that

$$G_+ - G_- \in \mathrm{Dom}(A) = \mathrm{Dom}(f(A)).$$

Therefore, the Livšic function $s_{(f(\dot{A}),f(A))}(z)$ has the representation

$$s_{(f(\dot{A}),f(A))}(z) = \frac{z-i}{z+i}\frac{(G_z,G_-)}{(G_z,G_+)}. \tag{E.7}$$

We have

$$(G_z,G_-) = \int_{\mathbb{R}} \frac{d\mu(\lambda)}{(\lambda - f^{-1}(z))\overline{(\lambda - f^{-1}(-i))}}$$

and

$$(G_z,G_+) = \int_{\mathbb{R}} \frac{d\mu(\lambda)}{(\lambda - f^{-1}(z))\overline{(\lambda - f^{-1}(i))}},$$

where $\mu(d\lambda)$ is the measure from the representation

$$M(z) = \int_{\mathbb{R}} \left(\frac{1}{\lambda - z} - \frac{\lambda}{1+\lambda^2}\right) d\mu(\lambda), \quad \operatorname{Im}(z) \neq 0.$$

Therefore,

$$\begin{aligned}
(z-i)(G_z,G_-) &= (z-i)\int_{\mathbb{R}} \frac{d\mu(\lambda)}{(\lambda - f^{-1}(z))(\lambda - f^{-1}(i))} \\
&= \frac{z-i}{f^{-1}(i) - f^{-1}(z)}\left(M(f^{-1}(z)) - M(f^{-1}(i))\right) \\
&= -a\left(M(f^{-1}(z)) - M(f^{-1}(i))\right)
\end{aligned} \tag{E.8}$$

and

$$\begin{aligned}
(z+i)(G_z,G_+) &= (z+i)\int_{\mathbb{R}} \frac{d\mu(\lambda)}{(\lambda - f^{-1}(z))(\lambda - f^{-1}(-i))} \\
&= \frac{z+i}{f^{-1}(-i) - f^{-1}(z)}\left(M(f^{-1}(z)) - M(f^{-1}(-i))\right) \\
&= -a\left(M(f^{-1}(z)) - M(f^{-1}(-i))\right).
\end{aligned} \tag{E.9}$$

Now (E.7), (E.8) and (E.9) yield

$$s_{(f(\dot{A}),f(A))}(z) = \frac{M(f^{-1}(z)) - M(f^{-1}(i))}{M(f^{-1}(z)) - M(f^{-1}(-i))} = \frac{m(z) - m(i)}{m(z) - \overline{m(i)}},$$

which completes the proof. $\qquad\square$

Next, we discuss the transformation law under the affine transformation of the pair $(\dot{A}, A)$ given by

$$(\dot{A}, A) \longrightarrow (-\dot{A}, -A).$$

Lemma E.4. *If $\dot{A}$ is a closed symmetric operator with deficiency indices $(1,1)$ and A its self-adjoint extension, then the Weyl-Titchmarsh functions $M_\pm(z)$ associated with the pairs $(\pm\dot{A}, \pm A)$ are related as follows*

$$M_-(z) = -\overline{M_+(-\bar{z})}, \quad z \in \rho(A). \tag{E.10}$$

In particular, for the Lišic functions associated with the pairs $(\dot{A}, A)$ and $(-\dot{A}, -A)$ we have

$$s_{(-\dot{A},-A)}(z) = \overline{s_{(\dot{A},A)}(-\bar{z})}, \quad z \in \mathbb{C}_+. \tag{E.11}$$

Proof. Let n be a unit vector in $\mathrm{Ker}((\dot{A})^* - iI)$ and

$$m = (A - iI)(A + iI)^{-1}n. \tag{E.12}$$

Then $m \in \mathrm{Ker}((\dot{A})^* + iI) = \mathrm{Ker}((-\dot{A})^* - iI)$. By the definition of the Weyl-Titchmarsh function one obtains that

$$M_-(z) = \left((-Az + I)(-A - zI)^{-1}m, m \right).$$

Therefore,

$$
\begin{aligned}
-M_-(-\bar{z}) &= ((A\bar{z} + I)(A - \bar{z}I)^{-1}m, m) \\
&= ((A\bar{z} + I)(A - \bar{z}I)^{-1}(A - iI)(A + iI)^{-1}n, \\
&\quad (A - iI)(A + iI)^{-1}n) \\
&= ((A - iI)(A + iI)^{-1}(A\bar{z} + I)(A - \bar{z}I)^{-1}n, \\
&\quad (A - iI)(A + iI)^{-1}n) \\
&= ((A\bar{z} + I)(A - \bar{z}I)^{-1}n, n) = M_+(\bar{z}) = \overline{M_+(z)}. \tag{E.13}
\end{aligned}
$$

Here we have used the Schwarz symmetry principle for the Weyl-Titchmarsh function

$$M_+(z) = \overline{M_+(\bar{z})}, \quad z \in \rho(A),$$

and the observation that the Cayley transform $(A - iI)(A + iI)^{-1}$ is a unitary operator commuting with the operator A. Finally, (E.10) follows from (E.13) by the substitution $z \to -\bar{z}$.

To prove the last assertion, we use the relation (2.4) to conclude that

$$s_{(-\dot{A},-A)}(z) = \frac{M_-(z) - i}{M_-(z) + i} = \frac{-\overline{M_+(-\overline{z})} - i}{-\overline{M_+(-\overline{z})} + i} = \overline{s_{(\dot{A},A)}(-\overline{z})},$$

completing the proof. □

Appendix F

THE INVARIANCE PRINCIPLE

The main goal of this chapter is to establish an invariance principle for the characteristic function of a triple of operators under linear transformations of the operators from the triple.

Introduce the class $\mathfrak{D}(\mathcal{H})$ of maximal dissipative unbounded densely defined operators $\widehat{A}$, $(\widehat{A} \neq (\widehat{A})^*)$, in the Hilbert space $\mathcal{H}$ such that

$$\dot{A} = \widehat{A}|_{\text{Dom}(\widehat{A}) \cap \text{Dom}(\widehat{A}^*)}$$

is a densely defined symmetric operator with deficiency indices $(1,1)$. In this case,

$$\dot{A} \subset \widehat{A} \subset (\dot{A})^*$$

and therefore $\widehat{A}$ is automatically a quasi-selfadjoint extension of $\dot{A}$ (see, e.g., [86]).

If $f(z)$ is the affine transformation $f(z) = az + b$, introduce the triple $f(\mathfrak{A})$ as

$$f(\mathfrak{A}) = (f(\dot{A}), f(\widehat{A}), f(A)).$$

Theorem F.1. *Let* $\mathfrak{A} = (\dot{A}, \widehat{A}, A)$ *be a triple such that* $\widehat{A} \in \mathfrak{D}(\mathcal{H})$. *Suppose that* $f(z) = az + b$ *with* $a, b \in \mathbb{R}$, $a > 0$, *is an affine transformation.*

Let $M(z)$ *be the Weyl-Titchmarsh function associated with the pair* $(\dot{A}, A)$. *Then the von Neumann parameters* $\varkappa$ *and* $\varkappa'$ *of the triples* $\mathfrak{A}$ *and* $f(\mathfrak{A})$ *are related as*

$$\frac{1 + \varkappa}{1 - \varkappa} = \frac{m - \varkappa'\overline{m}}{i(1 - \varkappa')}, \tag{F.1}$$

where

$$m = M(f^{-1}(i)).$$

Moreover, the characteristic functions $S_{f(\mathfrak{A})}(z)$ and $S_{\mathfrak{A}}(z)$ are related as

$$S_{f(\mathfrak{A})}(f(z)) = \Theta_f S_{\mathfrak{A}}(z), \quad z \in \mathbb{C}_+, \tag{F.2}$$

where

$$\Theta_f = \left(\frac{1-\varkappa}{1-\overline{\varkappa}}\right)^{-1} \cdot \frac{1-\varkappa'}{1-\overline{\varkappa'}}.$$

is a unimodular factor. In particular, Θ_f continuously depends on f.

Proof. As in the proof of Lemma E.3, from the very beginning one can assume that $\dot{A}$ is a prime symmetric operator.

Let $\mu(d\lambda)$ denote the representing measure for the Weyl-Titchmarsh function $M(z)$. Without loss of generality (see Theorem C.1) one may assume that A is the multiplication operator by independent variable in $L^2(\mathbb{R}, d\mu)$ and $\dot{A}$ coincides with its restriction on

$$\mathrm{Dom}(\dot{A}) = \left\{ h \in \mathrm{Dom}(A) \,\middle|\, \int_{\mathbb{R}} h(\lambda) d\mu(\lambda) = 0 \right\}.$$

In this case, from Theorem D.2 (see (D.4)), we know that the functions

$$g_+(\lambda) = \frac{1}{\lambda - i} \quad \text{and} \quad g_-(\lambda) = \frac{1}{\lambda + i}$$

form a basis in the deficiency subspace,

$$g_\pm \in \mathrm{Ker}((\dot{A})^* \mp iI), \quad \|g_\pm\| = 1.$$

From (C.9) is also follows that

$$g_+ - g_- \in \mathrm{Dom}(A). \tag{F.3}$$

Clearly, the functions

$$G_\pm(\lambda) = \frac{1}{\lambda - f^{-1}(\pm i)}$$

have the properties

$$G_\pm \in \mathrm{Ker}((f(\dot{A})^*) \mp I), \quad \|G_+\| = \|G_-\|,$$

and

$$G_+ - G_- \in \mathrm{Dom}(A) = \mathrm{Dom}(f(A)). \tag{F.4}$$

From the definition of the von Neumann parameters $\varkappa, \varkappa' \in \mathbb{D}$ for the triples $\mathfrak{A}$ and $f(\mathfrak{A})$ it follows that

$$g_+ - \varkappa g_- \in \mathrm{Dom}(\widehat{A}) \tag{F.5}$$

and

$$G_+ - \varkappa' G_- \in \mathrm{Dom}(f(\widehat{A})) = \mathrm{Dom}(\widehat{A}). \tag{F.6}$$

Introduce the function

$$m(z) = M(f^{-1}(z)), \quad \mathrm{Im}(z) \neq 0. \tag{F.7}$$

In order to establish the relationship (F.1) between the von Neumann parameters, notice that

$$G_\pm - \left[m(\pm i) \frac{g_+ - g_-}{2i} + \frac{g_+ + g_-}{2} \right] \in \mathrm{Dom}(f(\dot{A})), \tag{F.8}$$

that is,

$$\int_{\mathbb{R}} \left(G_\pm(\lambda) - \left[m(\pm i) \frac{g_+(\lambda) - g_-(\lambda)}{2i} + \frac{g_+(\lambda) + g_-(\lambda)}{2} \right] \right) d\mu(\lambda) = 0.$$

Indeed, since

$$\frac{g_+(\lambda) - g_-(\lambda)}{2i} = \frac{1}{\lambda^2 + 1}, \quad \frac{g_+(\lambda) + g_-(\lambda)}{2} = \frac{\lambda}{\lambda^2 + 1},$$

and

$$G_\pm(\lambda) = \frac{1}{\lambda - f^{-1}(\pm i)},$$

one needs to verify the equality

$$\int_{\mathbb{R}} \left(\frac{1}{\lambda - f^{-1}(\pm i)} - \left[\frac{M(f^{-1}(\pm i))}{\lambda^2 + 1} + \frac{\lambda}{\lambda^2 + 1} \right] \right) d\mu(\lambda) = 0,$$

which simply follows from the observations that

$$\int_{\mathbb{R}} \frac{M(f^{-1}(\pm i))}{\lambda^2 + 1} d\mu(\lambda) = M(f^{-1}(\pm i))$$

and

$$\int_{\mathbb{R}} \left(\frac{1}{\lambda - f^{-1}(\pm i)} - \frac{\lambda}{\lambda^2 + 1} \right) d\mu(\lambda) = M(f^{-1}(\pm i)).$$

Combining (F.6) and (F.8) we get that

$$h = \left(m\frac{g_+ - g_-}{2i} + \frac{g_+ + g_-}{2}\right) - \varkappa'\left(\overline{m}\frac{g_+ - g_-}{2i} + \frac{g_+ + g_-}{2}\right)$$

$$= \frac{m - \varkappa'\overline{m} + i - i\varkappa'}{2i}g_+ + \frac{i - i\varkappa' - m + \varkappa'\overline{m}}{2i}g_- \in \mathrm{Dom}(\widehat{A}).$$

Therefore, in view of (F.5),

$$\varkappa = \frac{m - \varkappa'\overline{m} - i + i\varkappa'}{m - \varkappa'\overline{m} + i - i\varkappa'} = \frac{m - i - \varkappa'(\overline{m} - i)}{m + i - \varkappa'(\overline{m} + i)} \tag{F.9}$$

and (F.1) follows.

From (F.5) and (F.6) it follows that the characteristic function $S_{\mathfrak{A}}(z)$ associated with the triple $\mathfrak{A} = (\dot{A}, \widehat{A}, A)$ (see (2.6)) can be evaluated as

$$S_{\mathfrak{A}}(z) = \frac{s_{(\dot{A},A)}(z) - \varkappa}{\overline{\varkappa}s_{(\dot{A},A)}(z) - 1}.$$

Representing $s_{(\dot{A},A)}(z)$ via the Weyl-Titchmarsh function $M(z)$,

$$s_{(\dot{A},A)}(z) = \frac{M(z) - i}{M(z) + i},$$

one concludes that

$$S_{\mathfrak{A}}(z) = \frac{\frac{M(z)-i}{M(z)+i} - \varkappa}{\overline{\varkappa}\frac{M(z)-i}{M(z)+i} - 1}.$$

Therefore, taking into account (F.7), one obtains

$$S_{\mathfrak{A}}(f^{-1}(z)) = \frac{\frac{m(z)-i}{m(z)+i} - \varkappa}{\overline{\varkappa}\frac{m(z)-i}{m(z)+i} - 1}. \tag{F.10}$$

In a similar way, using that

$$G_+ - G_- \in \mathrm{Dom}(f(A)) \quad \text{and} \quad G_+ - \varkappa G_+ \in \mathrm{Dom}(f(\widehat{A})),$$

one also gets

$$S_{f(\mathfrak{A})}(z) = \frac{s_{(f(\dot{A}),f(A))}(z) - \varkappa'}{\overline{\varkappa'}s_{(f(\dot{A}),f(A))}(z) - 1}.$$

By Lemma E.3,

$$s_{(f(\dot{A}),f(A))}(z) = \frac{m(z) - m}{m(z) - \overline{m}},$$

so that

$$S_{f(\mathfrak{A})}(z) = \frac{\frac{m(z)-m}{m(z)-\overline{m}} - \varkappa'}{\overline{\varkappa'}\frac{m(z)-m}{m(z)-\overline{m}} - 1}. \tag{F.11}$$

From (F.10) one gets that

$$S_{\mathfrak{A}}(f^{-1}(z)) = \frac{\frac{m(z)-i}{m(z)+i} - \varkappa}{\overline{\varkappa}\frac{m(z)-i}{m(z)+i} - 1} = \frac{m(z) - i - \varkappa(m(z)+i)}{\overline{\varkappa}(m(z)-i) - (m(z)+i)}$$

$$= \frac{(1-\varkappa)m(z) - i(1+\varkappa)}{(\overline{\varkappa}-1)m(z) - i(1+\overline{\varkappa})} = \frac{1-\varkappa}{\overline{\varkappa}-1} \cdot \frac{m(z) - i\frac{1+\varkappa}{1-\varkappa}}{m(z) + i\frac{1+\overline{\varkappa}}{1-\overline{\varkappa}}}. \tag{F.12}$$

A similar computation for the right hand side of (F.11) yields

$$S_{f(\mathfrak{A})}(z) = \frac{\frac{m(z)-m}{m(z)-\overline{m}} - \varkappa'}{\overline{\varkappa'}\frac{m(z)-m}{m(z)-\overline{m}} - 1} = \frac{m(z) - m - \varkappa'(m(z)-\overline{m})}{\overline{\varkappa'}(m(z)-m) - (m(z)-\overline{m})}$$

$$= \frac{(1-\varkappa')m(z) - (m - \varkappa'\overline{m})}{(\overline{\varkappa'}-1)m(z) + (\overline{m} - \overline{\varkappa'}m)} = \frac{1-\varkappa'}{\overline{\varkappa'}-1} \cdot \frac{m(z) - \frac{m-\varkappa'\overline{m}}{1-\varkappa'}}{m(z) + \frac{\overline{m}-\overline{\varkappa'}m}{\overline{\varkappa'}-1}}$$

$$= \frac{1-\varkappa'}{\overline{\varkappa'}-1} \cdot \frac{m(z) - i\frac{1+\varkappa}{1-\varkappa}}{m(z) + i\frac{1+\overline{\varkappa}}{1-\overline{\varkappa}}}, \tag{F.13}$$

where we used the relation (F.1) on the last step.

Comparing (F.12) and (F.13), we obtain

$$S_{f(\mathfrak{A})}(f(z)) = \left(\frac{1-\varkappa}{1-\overline{\varkappa}}\right)^{-1} \cdot \frac{1-\varkappa'}{1-\overline{\varkappa'}} \cdot S_{\mathfrak{A}}(z),$$

which proves (F.2). □

We conclude this chapter by establishing an invariance principle under the anti-holomorphic transformation (involution) of the triple

$$\mathfrak{A} = (\dot{A}, \widehat{A}, A) \longrightarrow -\mathfrak{A}^* = (-\dot{A}, -(\widehat{A})^*, -A).$$

Theorem F.2. *Let $\dot{A}$ be a densely defined, closed symmetric operator with deficiency indices $(1,1)$, A its self-adjoint extension and $\widehat{A}$ quasi-selfadjoint dissipative extension of $\dot{A}$.*

Then the characteristic functions associated with the triples $\mathfrak{A} = (\dot{A}, \widehat{A}, A)$ and $-\mathfrak{A}^* = (-\dot{A}, -(\widehat{A})^*, -A)$ are related as follows

$$S_{-\mathfrak{A}^*}(z) = \overline{S_{\mathfrak{A}}(-\overline{z})}. \tag{F.14}$$

Proof. Let $g_\pm$ be normalized deficiency elements of $\dot{A}$,

$$g_\pm \in \mathrm{Ker}((\dot{A})^* \mp I),$$

such that

$$g_+ - g_- \in \mathrm{Dom}(A) \quad \text{and} \quad g_+ - \varkappa g_- \in \mathrm{Dom}(\widehat{A}).$$

Clearly,

$$g_\pm \in \mathrm{Ker}((-\dot{A})^* \pm I),$$

$$g_- - g_+ \in \mathrm{Dom}(A) = \mathrm{Dom}(-A),$$

and

$$g_- - \overline{\varkappa}g_+ \in \mathrm{Dom}((\widehat{A})^*) = \mathrm{Dom}((-\widehat{A})^*).$$

Hence, using Lemma E.4, one obtains

$$S_{-\mathfrak{A}^*}(z) = \frac{s_{(-\dot{A},-A)}(z) - \overline{\varkappa}}{\varkappa s_{(-\dot{A},-A)}(z) - 1} = \overline{\frac{s(\widehat{A},A)(-\overline{z}) - \overline{\varkappa}}{\varkappa s_{(\dot{A},A)}(-\overline{z}) - 1}}$$

$$= \overline{S_{\mathfrak{A}}(-\overline{z})}.$$

The proof is complete.   □

Appendix G

THE OPERATOR COUPLING AND THE MULTIPLICATION THEOREM

Introduce the class $\mathfrak{D}_0(\mathcal{H})$ of maximal dissipative densely defined operators $\widehat{A}$ in the Hilbert space $\mathcal{H}$ of the form

$$\widehat{A} = A + itP,$$

where $A = \operatorname{Re}(\widehat{A})$ is a self-adjoint operator, $t > 0$, and P is a rank-one orthogonal projection [4, 15, 77].

Introduce the concept of an operator coupling of two operators from the classes $\mathfrak{D}_0(\mathcal{H}_1)$ and $\mathfrak{D}_0(\mathcal{H}_2)$.

Definition G.1. Suppose that $\widehat{A}_1 \in \mathfrak{D}_0(\mathcal{H}_1)$ and $\widehat{A}_2 \in \mathfrak{D}_0(\mathcal{H}_2)$ are maximal dissipative operators acting in the Hilbert spaces $\mathcal{H}_1$ and $\mathcal{H}_2$, respectively.

We say that a maximal dissipative operator $\widehat{A}$ from the class $\mathfrak{D}_0(\mathcal{H}_1 \oplus \mathcal{H}_2)$ is an operator coupling of $\widehat{A}_1$ and $\widehat{A}_2$, in writing,

$$\widehat{A} = \widehat{A}_1 \uplus \widehat{A}_2,$$

if $\widehat{A} - (\widehat{A}_1 \oplus \widehat{A}_2)$ is a rank-one operator, the Hilbert space $\mathcal{H}_1$ is invariant for $\widehat{A}$, and the restriction of $\widehat{A}$ on $\mathcal{H}_1$ coincides with the dissipative operator $\widehat{A}_1$. That is,

$$\operatorname{Dom}(\widehat{A}) \cap \mathcal{H}_1 = \operatorname{Dom}(\widehat{A}_1)$$

and

$$\widehat{A}\big|_{\mathcal{H}_1 \cap \operatorname{Dom}(\widehat{A}_1)} = \widehat{A}_1.$$

Theorem G.2 (cf. [4, 14, 15, 82]). *Let $\widehat{A} = \widehat{A}_1 \uplus \widehat{A}_2$ be an operator coupling of two maximal dissipative operators $\widehat{A}_k \in \mathfrak{D}_0(\mathcal{H}_k)$, $k = 1, 2$. Then the characteristic function of an operator coupling $\widehat{A}_1 \uplus \widehat{A}_2$ coincides with the product of the ones of $\widehat{A}_1$ and $\widehat{A}_2$,*

$$S_{\widehat{A}_1 \uplus \widehat{A}_2}(z) = S_{\widehat{A}_1}(z) \cdot S_{\widehat{A}_2}(z), \quad z \in \mathbb{C}_+. \tag{G.1}$$

Proof. Suppose that

$$\widehat{A}_k = A_k + i(\cdot, g_k)g_k, \quad k = 1, 2,$$

where $A_k = A_k^*$, $g_k \in \mathcal{H}_k$. Denote by P_k $(k = 1, 2)$ the orthogonal projections onto the subspaces $\mathcal{H}_k$, respectively. From the definition of an operator coupling it follows that

$$\widehat{A} = \widehat{A}_1 P_1 + \widehat{A}_2 P_2 + (\cdot, \widetilde{g})g$$

for some $g, \widetilde{g} \in \mathcal{H}_1 \oplus \mathcal{H}_2$ and that

$$\widehat{A} P_1 = \widehat{A}_1 P_1.$$

In particular,

$$(\widehat{A})^* P_2 = (\widehat{A}_2)^* P_2$$

and therefore

$$\widetilde{g} \in \mathcal{H}_2 \quad \text{and} \quad g \in \mathcal{H}_1. \tag{G.2}$$

First we show that

$$\text{Im}(\widehat{A}) = (\cdot, \phi)\phi,$$

where

$$\phi = (\Theta_1 g_1) \oplus (\Theta_2 g_2) \tag{G.3}$$

for some $|\Theta_k| = 1$, $k = 1, 2$. Indeed, we have

$$(\cdot, g_1)g_1 + (\cdot, g_2)g_2 + \frac{1}{2i}((\cdot, \widetilde{g})g - (\cdot, g)\widetilde{g}) = (\cdot, \phi)\phi. \tag{G.4}$$

Introducing

$$\phi_k = P_k \phi, \quad k = 1, 2,$$

from (G.2) and (G.4) it follows that

$$|(g_k, g_k)| = |(g_k, \phi)| = |(g_k, \phi_k)|, \quad k = 1, 2,$$

and then we get (G.3).

Rewrite the equality (G.4) one more time

$$(\cdot, g_1)g_1 + (\cdot, g_2)g_2 + \frac{1}{2i}\left((\cdot, \widetilde{g})g - (\cdot, g)\widetilde{g}\right)$$

$$= (\cdot, (\Theta_1 g_1) \oplus (\Theta_2 g_2))(\Theta_1 g_1) \oplus (\Theta_2 g_2).$$

We get

$$\frac{1}{2i}\left((\cdot, \widetilde{g})g - (\cdot, g)\widetilde{g}\right) = \overline{\Theta_1}\Theta_2(\cdot, g_1)g_2 + \Theta_1\overline{\Theta_2}(\cdot, g_2)g_1$$

and therefore

$$(\cdot, \widetilde{g})g = 2i(\cdot, \phi_2)\phi_1.$$

In particular, we have that

$$\widehat{A} = (A_1 + i(\cdot, \phi_1)\phi_1)P_1 + (A_2 + i(\cdot, \phi_2)\phi_2)P_2 + 2i(\cdot, \phi_2)\phi_1,$$

$$\mathrm{Im}(\widehat{A}) = (\cdot, (\phi_1 + \phi_2))(\phi_1 + \phi_2)$$

and we arrive at the definition of an operator coupling as presented in [16, eq. (2.1)]. Literally repeating step by step the proof of the Multiplication Theorem [15, Theorem 2.1] one justifies (G.1). □

Remark G.3. We remark that an operator coupling of two dissipative operators from the classes $\mathfrak{D}_0(\mathcal{H}_1)$ and $\mathfrak{D}_0(\mathcal{H}_2)$ is not unique. In fact, we have shown that an operator coupling $\widehat{A}_1 \uplus \widehat{A}_2$ of two dissipative operators

$$\widehat{A}_k = A_k + i(\cdot, g_k)g_k, \quad k = 1, 2,$$

is necessarily of the form

$$\widehat{A}_1 \uplus \widehat{A}_2 = (A_1 + i(\cdot, g_1)g_1)P_1 + (A_2 + i(\cdot, g_2)g_2)P_2 + 2i\Theta(\cdot, g_2)g_1, \quad (\text{G.5})$$

for some $|\Theta|=1$. Moreover, for any choice of Θ such that $|\Theta| = 1$ the right hand side of (G.5) meets the requirements to be an operator coupling of $\widehat{A}_1$ and $\widehat{A}_2$.

Recall that the class $\mathfrak{D}(\mathcal{H})$ consists of all maximal dissipative unbounded densely defined operators $\widehat{A}$, $(\widehat{A} \neq (\widehat{A})^*)$, in the Hilbert space $\mathcal{H}$ such that

$$\dot{A} = \widehat{A}\big|_{\mathrm{Dom}(\widehat{A})\cap\mathrm{Dom}((\widehat{A})^*)}$$

is a densely defined symmetric operator with deficiency indices $(1, 1)$ (see Appendix F).

Definition G.4 ([87]). Suppose that $\widehat{A}_1 \in \mathfrak{D}(\mathcal{H}_1)$ and $\widehat{A}_2 \in \mathfrak{D}(\mathcal{H}_2)$. We say that a $\widehat{A} \in \mathfrak{D}(\mathcal{H}_1 \oplus \mathcal{H}_2)$ is an operator coupling of $\widehat{A}_1$ and $\widehat{A}_2$, in writing,

$$\widehat{A} = \widehat{A}_1 \uplus \widehat{A}_2,$$

if

(i) the Hilbert space $\mathcal{H}_1$ is invariant for $\widehat{A}$ and the restriction of $\widehat{A}$ on $\mathcal{H}_1$ coincides with the dissipative operator $\widehat{A}_1$, that is,

$$\mathrm{Dom}(\widehat{A}) \cap \mathcal{H}_1 = \mathrm{Dom}(\widehat{A}_1),$$

$$\widehat{A}\big|_{\mathcal{H}_1 \cap \mathrm{Dom}(\widehat{A}_1)} = \widehat{A}_1,$$

and

(ii) the symmetric operator $\dot{A} = \widehat{A}\big|_{\mathrm{Dom}(\widehat{A}) \cap \mathrm{Dom}((\widehat{A})^*)}$ has the property

$$\dot{A} \subset \widehat{A}_1 \oplus (\widehat{A}_2)^*.$$

The corresponding multiplication theorem for the class $\mathfrak{D}(\mathcal{H})$ can be formulated as follows (see [87, Theorem 6.1, cf. Theorem 5.4]).

Theorem G.5. *Suppose that $\widehat{A} = \widehat{A}_1 \uplus \widehat{A}_2$ is an operator coupling of two maximal dissipative operators $\widehat{A}_k \in \mathfrak{D}(\mathcal{H}_k)$, $k = 1, 2$. Denote by $\dot{A}$, $\dot{A}_1$ and $\dot{A}_2$ the corresponding symmetric operators with deficiency indices $(1, 1)$, respectively. That is,*

$$\dot{A} = \widehat{A}\big|_{\mathrm{Dom}(\widehat{A}) \cap \mathrm{Dom}((\widehat{A})^*)}$$

and

$$\dot{A}_k = \widehat{A}_k\big|_{\mathrm{Dom}(\widehat{A}_k) \cap \mathrm{Dom}((\widehat{A}_k)^*)}, \quad k = 1, 2.$$

Then there exist self-adjoint reference operators A, A_1, and A_2, extending $\dot{A}$, $\dot{A}_1$ and $\dot{A}_2$, respectively, such that

$$S_{(\dot{A}, \widehat{A}_1 \uplus \widehat{A}_2, A)}(z) = S_{(\dot{A}_1, \widehat{A}_1, A_1)}(z) \cdot S_{(\dot{A}_2, \widehat{A}_2, A_2)}(z), \quad z \in \mathbb{C}_+. \tag{G.6}$$

Moreover, for any operator coupling $\widehat{A}$ of $\widehat{A}_1$ and $\widehat{A}_2$, the multiplication rule

$$\widehat{\kappa}(\widehat{A}) = \widehat{\kappa}(\widehat{A}_1) \cdot \widehat{\kappa}(\widehat{A}_2) \tag{G.7}$$

holds. Here $\widehat{\kappa}(\cdot)$ stands for the absolute value of the von Neumann parameter of a dissipative operator defined by (2.16).

Corollary G.6. *Assume the hypotheses of Theorem G.5. Then the von Neumann logarithmic potential $\Gamma_{\widehat{A}}(z)$ (see Definition 2.4) is an additive functional in the sense that*

$$\Gamma_{\widehat{A}_1 \uplus \widehat{A}_2}(z) = \Gamma_{\widehat{A}_1}(z) + \Gamma_{\widehat{A}_2}(z), \quad z \in \rho_{\widehat{A}_1} \cap \rho_{\widehat{A}_2} \cap \rho_{\widehat{A}_1 \uplus \widehat{A}_2} \cap \mathbb{C}_+.$$

Appendix H

STABLE LAWS

Recall (see, e.g., [33, 48, 141]) that a distribution G (of a random variable) is said to be stable, if a linear combination of two independent random variables with this distribution has the same distribution, up to location and scale parameters. That is, for any b_1, $b_2 > 0$, there exist a $b > 0$ and $a \in \mathbb{R}$ such that

$$G\left(\frac{x}{b_1}\right) \star G\left(\frac{x}{b_2}\right) = G\left(\frac{x-a}{b}\right),$$

where $\star$ denotes the convolution of distributions (see [33, Ch. V.4]).

It turns out that a (non-degenerated) law G is stable if and only if the logarithm of its characteristic function has the representation [141, Theorem B.2]

$$\log g(t) = \sigma\left(it\gamma - |t|^\alpha\left(1 - i\beta\frac{t}{|t|}\omega(t,\alpha)\right)\right) \tag{H.1}$$

for some $\sigma > 0$, $-\infty < \gamma < \infty$,

$$0 < \alpha \le 2, \quad \text{(the index of stability)}$$

$$-1 \le \beta \le 1, \quad \text{(the skew parameter).} \tag{H.2}$$

Here

$$\omega(t,\alpha) = \begin{cases} \tan\left(\frac{\pi}{2}\alpha\right), & \alpha \ne 1, \\ -\frac{2}{\pi}\log|t|, & \alpha = 1. \end{cases} \tag{H.3}$$

The skew parameter β is irrelevant when $\alpha = 2$.

Recall (see, e.g., [141]) that a distribution F is said to belong to *the domain of attraction of a law* if there are constants A_n and $B_n > 0$ such that the following non-zero limit

$$\lim_{n \to \infty} \log \left[f(t/B_n) \right]^n e^{iA_n t}$$

exists, where $f(t)$ is the characteristic function of the distribution F,

$$f(t) = \int_{\mathbb{R}} e^{itx} dF(x).$$

In this case the limit coincides with the logarithm of a stable law (H.1) for an appropriate choice of the parameters α, β, γ and σ.

Recall that a positive function $h(x)$, defined for $x \geq 0$, is said to be slowly varying if, for all $t > 0$,

$$\lim_{x \to \infty} \frac{h(tx)}{h(x)} = 1.$$

Also, by the Karamata theorem (see, e.g., [48, Appendix 1] for an exposition of the Karamata theory), a slowly varying function h which is integrable on any finite interval can be represented in the form

$$h(x) = c(x) \exp \left\{ \int_{x_0}^{x} \frac{\varepsilon(t)}{t} dt \right\}, \quad x_0 > 0,$$

where

$$\lim_{x \to \infty} c(x) = c \neq 0 \quad \text{and} \quad \lim_{x \to \infty} \varepsilon(x) = 0.$$

A key result in this area is the following Gnedenko-Kolmogorov limit theorem.

Theorem H.1 ([48, Theorem 2.6.1]). *A distribution F belongs to the domain of attraction of a stable law (H.1) with exponent α, $0 < \alpha \leq 2$, and parameters σ, β and γ if and only if*

$$1 - F(x) = \frac{c_1 + o(1)}{x^\alpha} h(x), \quad x > 0 \tag{H.4}$$

$$F(x) = \frac{c_2 + o(1)}{(-x)^\alpha} h(-x), \quad x < 0 \tag{H.5}$$

as $|x| \to \infty$, where $c_1, c_2 \geq 0$, $c_1 + c_2 > 0$ and h is slowly varying in the sense of Karamata.

In this case,

$$\sigma = (c_1 + c_2)d(\alpha), \tag{H.6}$$

where

$$d(\alpha) = \begin{cases} \Gamma(1 - \alpha) \cos\left(\frac{1}{2}\pi\alpha\right), & \alpha \neq 1 \\ \frac{\pi}{2}, & \alpha = 1 \end{cases}, \tag{H.7}$$

and

$$\beta = \frac{c_1 - c_2}{c_1 + c_2}. \tag{H.8}$$

Remark H.2. The (tauberian type) relationship between the set of data (c_1, c_2, h) and $(\alpha, \beta, \gamma, \sigma)$ referred to in Theorem H.1 (also see (H.2)) can be described as follows: if a distribution F belongs to the domain of attraction of the stable law (H.1), that is,

$$\lim_{n \to \infty} \log\left[f(t/B_n)\right]^n e^{iA_n t} = \sigma\left(it\gamma - |t|^\alpha + i\beta\frac{t}{|t|}\omega(t, \alpha)\right)$$

for some constants A_n and $B_n > 0$, then (see, e.g., [48, Theorem 2.6.5])

$$\log f(t) = i\tilde{\gamma}t - \sigma|t|^\alpha h(1/t)\left(1 - i\beta\frac{t}{|t|}\omega(t, \alpha)\right)(1 + o(1)) \quad \text{as} \quad t \to 0,$$

where $\tilde{\gamma}$ is in general not necessarily the same as γ. Recall that in this case the norming constants B_n necessarily satisfy the relation

$$\lim_{n \to \infty} nB_n^{-\alpha}h(B_n) = 1.$$

If in the hypothesis of Theorem H.1 the slowly varying function $h(x)$ has the property that $\lim_{x \to \infty} h(x) = 1$, then the scaling factors B_n can be given by

$$B_n = n^{1/\alpha}.$$

Under this hypothesis, the probability distribution F is said to belong to the domain of *normal* attraction of a stable law. In particular, every stable law belongs to the normal of its own normal attraction.

REFERENCES

[1] V. M. Adamjan, D. Z. Arov, *Unitary couplings of semi-unitary operators,* Mat. Issled. vyp. 2, 3–64 (1966).

[2] Y. Aharonov, D. Bohm, *Significance of Electromagnetic Potentials in the Quantum Theory,* Phys. Rev. **115**, 485–491 (1959).

[3] N. I. Akhiezer and I. M. Glazman, Theory of Linear Operators in Hilbert Space, Dover, New York, 1993.

[4] Yu. Arlinskii, S. Belyi, E. Tsekanovskii, Conservative realizations of Herglotz-Nevanlinna functions, Operator Theory: Advances and Applications **217**, Birkhäuser, 2011.

[5] Yu. M. Arlinskii, V. A. Derkach, E. R. Tsekanovskii, *On unitarily equivalent quasi-Hermitian extensions of Hermitian operators,* (Russian), Math. Physics, Institute of Mathematics of the Ukrainian Academy of Sciences, Respubl. Sb., Naukova Dumka, Kiev **29**, 71–77 (1981).

[6] H. Atmanspacher, W. Ehm, T. Gneiting, *Necessary and sufficient conditions for the quantum Zeno and anti-Zeno effect,* J. Physics A: Math. Gen. **36**, 9899–9905 (2003).

[7] S. Belyi, E. Tsekanovskii, *Perturbations of Donoghue classes and inverse problems for L-systems,* Complex Analysis and Operator Theory **13**, 1227–1311 (2019).

[8] G. Berkolaiko, P. Kuchment, Introduction to Quantum Graphs, American Mathematical Society. Providence, Rhode Island, 2013.

[9] L. Bracci, L. E. Picasso, *Representations of semigroups of partial isometries,* Bull. Lond. Math. Soc. **39**, no. 5, 792–802 (2007).

[10] L. de Branges, J. Rovnyak, Canonical models in quantum scattering theory. 1966 Perturbation Theory and its Applications in Quantum Mechanics (Proc. Adv. Sem. Math. Res. Center, U.S. Army, Theoret. Chem. Inst., Univ. of Wisconsin, Madison, Wis., (1965) pp. 295–392, Wiley, New York.

[11] O. Bratteli, D. W. Robinson, Operator Algebras and Quantum Statistical Mechanics I. C^*- and W^*-Algebras. Symmetry Groups. Decomposition of States. Springer-Verlag, New York Heidelberg Berlin, 1979.

[12] O. Bratteli, D. W. Robinson, Operator Algebras and Quantum Statistical Mechanics II. Equilibrium States. Models in Quantum Statistical Mechanics. Springer-Verlag, New York Heidelberg Berlin, 1981.

[13] M. S. Brodskii and M. S. Livšic, *On linear operator-valued functions invariant with respect to the translation group*, (Russian), Doklady Akad. Nauk SSSR (N.S.) **68**, 213–216 (1949).

[14] M. S. Brodskii and M. S. Livšic, *Spectral analysis of non-selfadjoint operators and intermediate systems*. Amer. Math. Soc. Transl. (2) **13**, 265–346 (1960).

[15] M. S. Brodskii, Triangular and Jordan Representations of linear operators. American Mathematical Society Providence, Rhode Island, 1971.

[16] L. E. J. Brouwer, *Beweis der Invarianz des n-dimensionalen Gebiets*, Mathematische Annalen **71**, 305–315 (1912).

[17] E. B. Davies, Quantum theory of open systems. London, New York, Academic Press, 1976.

[18] V. A. Derkach, M. M. Malamud, *Generalized resolvents and the boundary value problems for Hermitian operators with gaps*. J. Funct. Anal. **95**, 1–95 (1991).

[19] V. A. Derkach, M. M. Malamud, Extension theory of symmetric operators and boundary value problems, (Russian), Proceedings of the Institute of Mathematics of NAS of Ukraine v. **104**, 2017.

[20] D. P. Di Vincenzo, E. J. Mele, *Self-consistent effective-mass theory for intralayer screening in graphite intercalation compounds*, Phys. Rev. B **29**, 1685–1694 (1984).

[21] J. Dixmier, *Sur la relation $i(PQ - QP) = 1$*. (French) Compositio Math. **13**, 263–269 (1958).

[22] G. Doetsch, Introduction to the Theory and Application of the Laplace Transformation, Springer-Verlag, 1974.

[23] W. F. Donoghue, *On perturbation of spectra*, Commun. Pure and Appl. Math. **18**, 559–579 (1965).

[24] V. K. Dubovoi, *The Weyl family of operator colligations and the corresponding open fields*, (Russian), Theory of Func., Funct. Anal. Appl. **14**, Kharkov Univ. Press, (1971).

[25] V. K. Dubovoi, *Invariant operator colligations*, (Russian), Vestnik Har'kov. Gos. Univ. No. **67**, 36–61 (1971).

[26] V. Efimov, *Energy levels arising from resonant two-body forces in a three-body system*, Phys. Lett. B **33**, 563–564 (1970).

[27] G. G. Emch, Algebraic Methods in Statistical Mechanics and Quantum Field Theory, Interscience Monographs and Texts in Physics and Astronomy, **XXVI**, Wiley & Sons, Inc., 1972.

[28] A. C. Elitzur, L. Vaidman, *Quantum Mechanical Interaction-Free Measurements*, Foundations of Physics **23**, No. 7, 987–997 (1993).

[29] P. Exner, Open Quantum systems and Feynman Integrals, Fundamental Theories of Physics. D. Reidel Publishing Co., Dordrecht, 1985.

[30] P. Exner, *Sufficient conditions for the anti-Zeno effect*, J. Phys. **A 38**, no. 24, L449–L454 (2005).

[31] L. D. Faddeev, S. P. Merkuriev, Quantum Scattering Theory for Several Particle Systems, Kluwer Acad. Publ., Dordrecht, 1993.

[32] W. Feller, An introduction to Probability Theory and its Applications, Vol. 1, John Wiley &Sons, Inc., New York-London-Sydney, 1957.

[33] W. Feller, An introduction to Probability Theory and its Applications, Vol. 2, John Wiley &Sons, Inc., New York-London-Sydney, 1966.

[34] F. P. Feynman, QED: The strange theory of light and matter. Princeton University Press, 2006.

[35] F. P. Feynman, A. R. Hibbs, Quantum Mechanics and Path Integrals. New York: McGraw-Hill, 1965.

[36] V. A. Fock, V. A. Krylov, Journal of Experimental and Theoretical Physics (USSR) **17**, 93 (1947).

[37] G. B. Folland, Real Analysis: Modern Techniques and Their Applications. Second Edition. John Wiley & Sons, Inc., 1999.

[38] C. Foias, L. Gehèr, B. Sz.-Nagy, *On the permutability condition of quantum mechanics*, Acta Sci. Math. (Szeged) **21**, 78–89 (1960).

[39] G. Gamow, *Zur Quantentheorie des Atomkernes,* Zeitschrift für Physik **51**, 204–212 (1928).

[40] J. S. Geronimo, T. P. Hill, *Necessary and sufficient condition that the limit of Stieltjes transforms is a Stieltjes transform,* J. of Appr. Theory **121**, 54–60 (2003).

[41] F. Gesztesy, N. Kalton, K. A. Makarov, E. Tsekanovskii, *Some applications of operator-valued Herglotz functions.* Operator theory, system theory and related topics (Beer-Sheva/Rehovot, 1997), 271–321, Oper. Theory Adv. Appl., **123**, Birkhäuser, Basel, 2001.

[42] F. Gesztesy, K. A. Makarov, E. Tsekanovskii, *An addendum to Krein's formula,* J. Math. Anal. Appl. **222**, 594–606 (1998).

[43] F. Gesztesy, E. Tsekanovskii, *On Matrix-Valued Herglotz Functions,* Math. Nachr. **218**, 61–138 (2000).

[44] W. Heisenberg, The physical principles of the quantum theory. The University of Chicago Press, Chicago, Illinois, 1930.

[45] G. C. Hegerfeldt, *Causality, particle localization and positivity of the energy.* In: Irreversibility and causality: semigroups and rigged Hilbert spaces. Lecture Notes in Physics v. **504**, 238–245 (1998).

[46] A. S. Holevo, Quantum Systems, Channels, Information: A Mathematical Introduction (Texts and Monographs in Theoretical Physics), de Gruyter GmbH, Walter, 2019.

[47] J. Holtsmark, *Über die Verbreiterung von Spektrallinien,* Annalen der Physik, B. **58**, 577–630 (1919).

[48] I. A. Ibragimov and Ju. V. Linnik, Independent and stationary sequences of random variables, Wolters-Noordhoff Publishing Groningen The Netherlands, 1971.

[49] J. M. Jauch, Foundations of Quantum Mechanics, Adison-Wesley Publishing Company, 1968.

[50] P. E. T. Jørgensen, *Selfadjoint operator extensions satisfying the Weyl commutation relations*, Bull. Amer. Math. Soc. (N.S.) **1**, no, 1, 266–269 (1979).

[51] P. E. T. Jørgensen, *Selfadjoint extension operators commuting with an algebra.* Math. Z. **169**, no. 1, 41–62 (1979).

[52] P. E. T. Jørgensen, P. S. Muhly, *Selfadjoint extensions satisfying the Weyl operator commutation relations,* J. Analyse Math. **37**, 46–99 (1980).

[53] P. E. T. Jørgensen, *Commutators of Hamiltonian operators and nonabelian algebras,* J. Math. Anal. Appl. **73**, no. 1, 115–133 (1980).

[54] P. E. T. Jørgensen, *A uniqueness theorem for the Heisenberg-Weyl commutation relations with nonselfadjoint position operator,* Amer. J. Math. **103**, no. 2, 273–287 (1981).

[55] P. E. T. Jørgensen, R. T. Moore, Operator Commutation Relations: Commutation Relations For Operators, Semigroups, And Resolvents With Applications To Mathematical Physics And Representations Of Lie Dordrecht; Boston: D. Reidel Pub. Co.; Hingham, MA: Sold and distributed in the U.S.A. and Canada by Kluwer Academic Publishers, 1984.

[56] I. S. Kac, M. G. Krein, *R-functions-analytic functions mapping upper halfplane into itself,* Amer. Math. Soc. Transl. (2) **103**, 1–18 (1974).

[57] T. Kato, *On the commutation relation $AB - BA = c$,* Arch. Rational Mech. Anal. **10** 273–275 (1962).

[58] L. A. Khalfin, *On the theory of the decay of a quasi-stationary state,* Dokl. Akad. Nauk SSSR, 1957, **115**, Number 2, 277–280. (Russian). English translation in: Soviet Phys. Doklady, **2**, 340–343 (1958).

[59] V. Kostrykin, R. Schrader, *Quantum wires with magnetic fluxes,* Dedicated to Rudolf Haag. Comm. Math. Phys. **237**, 161–179 (2003).

[60] P. Koosis, Introduction to H_p spaces. Second edition. With two appendices by V. P. Havin [Viktor Petrovich Khavin]. Cambridge Tracts in Mathematics, **115**. Cambridge University Press, Cambridge, 1998.

[61] M. A. Krasnosel'skii, *On self-adjoint extensions of Hermitian operators,* (Russian), Ukrain. Mat. Zurnal **1**, 21–38 (1949).

[62] M. G. Krein, *On resolvent of Hermitian operator with defect numbers (m, m),* (Russian), Dokl. Akad. Mauk SSSR **52**, no. 8, 657–660 (1946).

[63] M. G. Krein, *The theory of self-adjoint extensions of semi-bounded Hermitian transformations and its applications. I,* (Russian), Rec. Math. [Mat. Sbornik] N.S. **20(62)**, 431–495 (1947).

[64] M. G. Krein, H. Langer, *Über die Q-Funktion eines π-hermiteschen Operators im Raume $\Pi_\varkappa$,* (German), Acta Sci. Math. (Szeged) **34**, 191–230, (1973).

[65] M.G. Krein, I. E. Ovčarenko, *Inverse problems for Q-functions and resolvent matrices of positive Hermitian operators,* (Russian), Dokl. Akad. Nauk SSSR **242**, no 3, 521–524 (1978).

[66] V. Kruglov, K. A. Makarov, B. Pavlov, A. Yafyasov, *Exponential decay in quantum mechanics.* Computation, physics and beyond, 268–288, Lecture Notes in Comput. Sci., 7160, Springer, Heidelberg, 2012.

[67] P. Kurasov, *Inverse problems for Aharonov-Bohm rings.* Math. Proc. Cambridge Philos. Soc. **148**, 331–362 (2010).

[68] A. Kuzhel, Characteristic Functions and Models of Nonself-Adjoint Operators. Mathematics and its Applications, **349**. Kluwer Academic Publishers Group, Dordrecht, 1996.

[69] A. V. Kuzhel, S. A. Kuzhel, Regular extensions of Hermitian operators. VSP, Utrecht, 1998.

[70] P. Kwiat, H. Weinfurter, A. Zeilinger, *Quantum Seeing in the Dark.* Scientific American **275**, issue 5, 72–78 (1996).

[71] L. D. Landau and E. M. Lifshitz, Quantum Mechanics. Non-relativistic Theory. Pergamon Press. Oxford, 1965.

[72] P. D. Lax and R. S. Phillips, Scattering Theory, Second edition. With appendices by Cathleen S. Morawetz and Georg Schmidt. Pure and Applied Mathematics, 26. Academic Press, Inc., Boston, MA, 1989.

[73] M. S. Livšic, *On a class of linear operators in Hilbert space,* Mat. Sbornik (2) **19**, 239–262 (1946).

[74] G. Lindblad, Nonequilibrium entropy and irreversibility. Mathematical Physics Studies, 5. D. Reidel Publishing Co., Dordrecht, 1983.

[75] K. Lindenberg, B. J. West, The nonequilibrium statistical mechanics of open and closed systems. VCH Publishers, Inc., New York, 1990.

[76] W. von ver Linden, V. Dose, U. von Toussait, Bayesian Probability Theory, Applications in the Physical Sciences, Cambridge University Press, 2014.

[77] M. Livšic, *On spectral decomposition of linear non-self-adjoint operators.* Mat. Sbornik (76) **34** (1954), 145–198 (Russian); English translation in Amer. Math. Soc. Transl. (2) **5**, 67–114 (1957).

[78] M. S. Livšic, *The application of non-self-adjoint operators to scattering theory,* Soviet Physics JETP **4**, 91–98 (1957).

[79] M. S. Livšic, Operators, oscillations, waves (open systems). Translated from the Russian by Scripta Technica, Ltd. English translation edited by R. Herden. Translations of Mathematical Monographs, Vol. 34. American Mathematical Society, Providence, R.I., 1973.

[80] M. S. Livšic, N. Kravitsky, A. S. Markus, V. Vinnikov, Theory of Commuting Nonselfadjoint Operators. Mathematics and its Applications, 332. Kluwer Academic Publishers Group, Dordrecht, 1995.

[81] M. Livšic, A. Jantsevich, Theory of Operator Colligations in Hilbert Spaces, (Russian), Kharkov Univ. Press, 1971.

[82] M. S. Livšic, V. P. Potapov, *A theorem on the multiplication of characteristic matrix-functions,* (Russian), Dokl. Acad. Nauk SSSR **72**, 625–628 (1950).

[83] G. W. Mackey, *A theorem of Stone and von Neumann,* Duke Math. J. **16**, 313–326 (1949).

[84] K. A. Makarov, *Semiboundedness of the energy operator of a system of three particles with paired interactions of δ-function type.* (Russian) Algebra i Analiz, **4**, no. 5, 155–171 (1992); translation in St. Petersburg Math. J. **4**, no. 5, 967–980 (1993).

[85] K. A. Makarov and V. V. Melezhik, *Two sides of a coin: The Efimov effect and the collapse in the three-body system with point interactions. I*, Teoret. Mat. Fiz. **107**, 415–432 (1996); English translation in Theoret. and Math. Phys. **107**, 75–769 (1997).

[86] K. A. Makarov, E. Tsekanovskii, *On the Weyl-Titchmarsh and Livšic functions*, Proceedings of Symposia in Pure Mathematics, Vol. 87, 291–313, American Mathematical Society, 2013.

[87] K. A. Makarov, E. Tsekanovskii, *On the addition and multiplication theorem.* Recent advances in inverse scattering, Schur analysis and stochastic processes, 315–339, Oper. Theory Adv. Appl., **244**, Linear Oper. Linear Syst., Birkhäuser/Springer, Cham, 2015.

[88] K. A. Makarov, E. Tsekanovskii, *On dissipative and non-unitary solutions to operator commutation relations*, Teoret. Mat. Fiz. **186**, 51–75 (2016); English translation in Theoret. and Math. Phys. **186**, 43–62 (2016).

[89] K. A. Makarov, E. Tsekanovskii, *On μ-scale invariant operators*, Methods of Func. Anal. and Topology **113**, no. 2, 23–29 (2007).

[90] L. Mandelstam, Ig. Tamm, *The Uncertainty Relation Between Energy and Time in Non-relativistic Quantum Mechanics*, J. Phys. USSR **9**, 249–254 (1945) and I.E. Tamm Selected Papers, Eds. B. M. Bolotovskii and V. Ya. Frenkel, (Springer, Berlin,1991).

[91] G. Menon, S. Belyi, *Dirac particle in a box, and relativistic quantum Zeno dynamics*, Phys. Lett. **A 330**, no. 1–2, 33–40 (2004).

[92] R. A. Minlos, L. D. Faddeev, *On the point interaction for a three-particle system in quantum mechanics*, Dokl. Akad. Nauk SSSR **141**, 1335–1338 (1962); English translation in Soviet Physics Dokl. **6**, 1072–1074 (1962).

[93] B. Misra, E. C. G. Sudarshan, *The Zeno paradox in quantum theory*, J. Math. Phys. **18**, 756–763 (1977).

[94] S. N. Naboko, *Functional model of perturbation theory and its applications to scattering theory*, (Russian), Boundary value problems of mathematical physics, 10. Trudy Mat. Inst. Steklov. **147**, 86–114 (1980).

[95] B. Sz.-Nagy, C. Foias, Harmonic Analysis Of Operators On Hilbert Space. New York, Springer, 2010.

[96] N. K. Nikolski, Operators, Functions, and Systems: An Easy Reading: Volume 1: Hardy, Hankel, and Toeplitz. Translated from the French by Andreas Hartmann. Mathematical Surveys and Monographs, 92. American Mathematical Society, Providence, RI, 2002.

[97] N. K. Nikolski, Operators, Functions, and Systems: An Easy Reading: Volume 2: Model operators and systems. Translated from the French by Andreas Hartmann and revised by the author. Mathematical Surveys and Monographs, 93. American Mathematical Society, Providence, RI, 2002.

[98] J. von Neumann, *Die Eindeutigkeit der Schrödingerschen Operatoren*, (German), Math. Ann. **104**, 570–578 (1931).

[99] B. S. Pavlov, *Conditions for separation of the spectral components of a dissipative operator*, (Russian), Izv. Akad. Nauk SSSR Ser. Mat. **39**, 123–148 (1975).

[100] B. S. Pavlov, *Dilation theory and spectral analysis of nonselfadjoint differential operators*, (Russian), Mathematical programming and related questions (Proc. Seventh Winter School, Drogobych, 1974), Theory of operators in linear spaces (Russian), 3–69. Central. Ekonom. Mat. Inst. Akad. Nauk SSSR, Moscow (1976). English translation in Amer. Math. Soc. Tranl. (2) **115**, 103–142 (1980).

[101] B. S. Pavlov, *Selfadjoint dilation of a dissipative Schrödinger operator, and expansion in its eigenfunction*, (Russian), Mat. Sb. (N.S.) **102** (144), no. 4, 511–536 (1977).

[102] B. S. Pavlov, *The theory of extensions, and explicitly solvable models*, (Russian), Uspekhi Mat. Nauk **42**, no. 6(258), 99–131 (1987); English translation in Mathematical Surveys **42**:6, 127–168 (1987).

[103] B. S. Pavlov, *Boundary conditions on thin manifolds and the semiboundedness of the three-body Schrödinger operator with point potential*, (Russian), Mat. Sb. (N.S.) **136** (**178**), no. 2, 16–177 (1988); English translation in Math. USSR-Sb. **64**, no. 1, 161–175 (1989).

[104] B. S. Pavlov, S. I. Fedorov, *Shift group and harmonic analysis on a Riemann surface of genus one*, (Russian), Algebra i Analiz **1**, 132–168 (1989); English translation in Leningrad Math. J. **1**, 447–490 (1990).

[105] B. S. Pavlov, *Spectral analysis of a dissipative singular Schödinger operator in terms of a functional model*; Partial differential equations, VIII, 87–153, Encyclopaedia Math. Sci., 65, Springer, Berlin, 1996.

[106] B. S. Pavlov, *The spectral nature of resonances. I.* (Russian), Mathematics of the 20th century. A view from St. Petersburg, pp. 125–151, (A. M. Vershik ed.) Moscow Center for Continuous Mathematical Education, 2010.

[107] V. V. Peller, Hankel Operators and Their Applications. Springer Monographs in Mathematics. Springer-Verlag, New York, 2003.

[108] A. M. Perelomov, V. S. Popov, *Collapse onto scattering center in quantum mechanics*, (Russian), Theor. Mat. Fiz. **4**, no. 1, 48–65 (1970).

[109] A. M. Perelomov, Y. B. Zel'dovich, Quantum mechanics. Selected topics. World Scientific Publishing Co., Inc., River Edge, NJ, 1998.

[110] P. Pfeifer, J. Fröhlich, *Generalized time-energy uncertainty relations and bounds on lifetimes of resonances*, Rev. Mod. Phys. **67**, 759–779 (1995).

[111] A. Philimonov, E. Tsekanovskii, *Automorphically invariant operator colligations and factorization of its characteristic operator-valued functions*, (Russian), Funct. Anal. i Prilozhen. **21**, no. 4, 94–95 (1987).

[112] R. S. Phillips, *On dissipative operators*, in "Lecture Series in Differential Equations," Vol. II (A. K. Aziz ed.) von Nostrand, 65–113 (1969).

[113] C. R. Putnam, Commutation properties of Hilbert space operators and related topics. Ergebnisse der Mathematik und ihrer Grenzgebiete, Band 36 Springer-Verlag New York, Inc., New York, 1967.

[114] F. Rellich, *Der Eindeutigkeitssatz für die Lösungen der quantenmechanischen Vertauschungsrelationen*, (German) Nachr. Akad. Wiss. Göttingen. Math.-Phys. Kl. Math.-Phys.-Chem. Abt. 1946, 107–115 (1946).

[115] J. Rosenberg, *A selective history of the Stone-von Neumann theorem. Operator algebras, quantization, and noncommutative geometry.* Contemp. Math. **365**, 331–353, Amer. Math. Soc., Providence, RI, 2004.

[116] V. Ryzhov, *Functional model of a class of non-selfadjoint extensions of symmetric operators.* Operator theory, analysis and mathematical physics, 117–158, Oper. Theory Adv. Appl. **174**, Birkhäuser, Basel, 2007.

[117] V. Ryzhov, *Functional model of a closed non-selfadjoint operator,* Integral Equations Operator Theory **60**, no. 4, 539–571 (2008).

[118] L. A. Sahnovič, *Dissipative operators with absolutely continuous spectrum,* (Russian), Trudy Moskov. Mat. Obsc. **19**, 211–270 (1968).

[119] A. U. Schmidt, *Mathematics of the quantum Zeno effect,* Mathematical physics research on the leading edge, 113–143, Nova Sci. Publ., Hauppauge, NY, 2004.

[120] F. Schäfer, I. Herrera, S. Cherukattil, C. Lovecchio, F. S. Cataliotti, F. Caruso, A. Smerzi, *Experimental realization of quantum zeno dynamics,* Nature Communications **5**, 3194 (2014).

[121] K. Schmüdgen, *On the Heisenberg commutation relation.* I, J. Funct. Anal. **50**, 8–49 (1983).

[122] K. Schmüdgen, *On the Heisenberg commutation relation.* II, Publ. RIMS, Kyoto Univ. **19**, 601–671 (1983).

[123] R. Schrader, *On the existence of a local Hamiltonian in the Galilean invariant Lee Model,* Comm. Math. Phys. **10**, no. 2, 155–178 (1968).

[124] B. Simon, *Spectral analysis of rank one perturbations and applications.* Mathematical quantum theory. II. Schrödinger operators (Vancouver, BC, 1993), 109–149, CRM Proc. Lecture Notes, 8, Amer. Math. Soc., Providence, RI, 1995.

[125] Ja. G. Sinaĭ, *Dynamical systems with countable Lebesgue spectrum, I,* (Russian), Izv. Akad. Nauk SSSR Ser. Mat. **25**, 899–924 (1961).

[126] Yu. G. Shondin, *On the three-particle problem with δ-potentials,* Teoret. Mat. Fiz. **51**, 181–191 (1982); English translation in Theoret. and Math. Phys. **51**, 434–441 (1982).

[127] M. Stone, *On one-parameter unitary groups in Hilbert space,* Ann. Math. **33**, 643–648 (1932).

[128] A. V. Straus, *Characteristic functions of linear operators,* (Russian), Izv. Akad. Nauk SSSR, Ser. Math. **24**, no. 1, 43–74 (1960).

[129] A. V. Shtraus, *On the extensions and the characteristic function of a symmetric operator,* Izv. Akad. Nauk SSR, Ser. Mat. **32**, 186–207 (1968).

[130] L. Susskind, A. Friedman, Quantum Mechanics: The Theoretical Minimum. New York: Basic Books, 2014.

[131] L. A. Takhtajan, Quantum Mechanics for Mathematicians (Graduate Studies in Mathematics Volume 95). American Mathematical Society. Providence, Rhode Island, 2008.

[132] C. Teuscher (ed.), Alan Turing: Life and Legacy of a Great Thinker. Springer-Verlag Berlin Heidelberg, 2004.

[133] E. Tsekanovskii, *Triangular models of unbounded accretive operators and regular factorization of their characteristic operator-functions*, (Russian), Dokl. Akad. Nauk SSSR **297**, no. 3, 552–556 (1987). English translation in Soviet Math. Dokl. **36**, no. 3, 512–515 (1988).

[134] E. Tsekanovskii, Yu. Shmuljan, *The theory of bi-extensions of operators in rigged Hilbert spaces. Unbounded operator colligations and characteristic functions.* Uspechi Mat. Nauk, **32**, no. 5, 64–124 (1977). English translation in Russ. Math. Surv. **32**, 73–131 (1977).

[135] E. Tsekanovskii, *Characteristic function and sectorial boundary value problems*, (Russian), Proceedings of the Institute of Mathematics, Siberian branch of RAS, Research on Geometry and Math. Analysis, Nauka **7**, 180–194 (1987).

[136] J. W. Vick, Homology theory. An introduction to algebraic topology. Academic Press, 1973.

[137] A. Weil, *Sur certains groupes d'opérateurs unitaires.* (French) Acta Math. **111**, 143–211 (1964).

[138] V. E. Weisskopf, E. P. Wigner, *Berechnung der natürlichen Linienbreite auf Grund der Diracschen Lichttheorie*, Zeitshrift für Physik **63**, 54–73 (1930).

[139] G. Weyl, *Quantenmechanik and Gruppentheorie*, (German), Zeits. für Physik **46**, 1–47 (1928).

[140] V. A. Zolotarev, *The Lax-Phillips scattering scheme on groups, and the functional model of a Lie algebra*, (Russian), Mat. Sb. **183**, no 5, 115–144 (1992); English translation in Russian Acad. Sci. Sb. Math. **76**, no. 1, 99–122 (1993).

[141] B. M. Zolotarev, One-dimensional Stable Distributions, Translation of Mathematical Monographs, v. 65, AMS, 1986.

[142] V. A. Zolotarev, Analytic methods in spectral representations of non-self-adjoint and non-unitary operators, (Russian), Kharkov National University, Kharkov, 2003.

INDEX